Desertification and Risk Analysis Using High and Medium Resolution Satellite Data

Results of the NATO Advanced Research Workshop on
Desertification and Risk Analysis Using High and Medium Satellite Data
Tozeur, Tunisia
9–19 December 2006

Library of Congress Control Number: 2008938014

ISBN 978-1-4020-8936-7 (PB)
ISBN 978-1-4020-8935-0 (HB)
ISBN 978-1-4020-8937-4 (e-book)

Published by Springer,
P.O. Box 17, 3300 AA Dordrecht, The Netherlands.

www.springer.com

Printed on acid-free paper

PREFACE

This work becomes with methodological rigor a part of the innovative proposals for the characterization of the areas at risk of desertification.

The complexity of the phenomenon of desertification, which involves extended surfaces in all continents, is one of the most alarming processes of the environmental degradation of our Planet and threatens the health and the living conditions of over a billion of persons.

The food crisis, in continuous increase, ask for the world of research to urgently supply reassuring solutions concerning the acquisition of indicators, which are easy to monitor and concur to control the phenomenon in order to fight its acceleration.

The constant but different combination that determines the diffusion of desertification in the territories, attributable mostly to climatic changes and the activities of the man, makes our job challenging and complicated, since it varies from region to region.

We are aware of all this and therefore we think that the methodological approach of survey of the data is an extremely important element to locate of the phenomenon and to monitor its course. This work, which is characterized for its multi-disciplinary approach, suggests solutions that we wish will quickly find concrete applications at international level.

Prof. Bruno Dettori

President CNLSD
(Comitato Nazionale per la Lotta alla Siccità e alla Desertificazione)

GENERAL ASPECTS OF DESERTIFICATION

ANTHROPIC ACTIONS AND DESERTIFICATION IN ALGERIA

MOHAMED TAHAR BENAZZOUZ AND LAMIA BOUREBOUNE
University Mentouri Constantine, Algeria
mtbenazzouz@yahoo.fr; lambourbon@yahoo.fr

Abstract

The recent generalization of the management of the Sahara desert margins and the high pression of the population causes a large degradation of the environment that appears through a phenomenon of salinity of soils, destruction of natural vegetation, wind blown accumulation areas which affects oasis, roads.

The research of the causes and the understanding of the process of desertification allows us to build a better evaluation of the real impact of this management on the environment. This research is based on field-work, the effects of aeolian dynamic, cartography and localization of wind blown accumulation, assessment and control of some managements. At this phase we can establish the impact of those actions on the environment through the progress of desertification and we try to find the best way to fighting desertification in Algeria. The great project of the green dam confirms the absence of a real knowledge of the mechanism of desertification. The support of this policy of management based on great works as roads, confirms the acceleration of the degradation of the environment that will be irreversible.

In Algeria, desertification risk is coming from the High Plains in the North where more than 20 millions of hectares of soils are highly exposed and vulnerable to desertification because the present massive destruction of vegetation and soils.

Keywords: Zibans range, Hodna, Algeria, Aeolian Morphogenesis, Holocene, Desertification, Morphoclimatic variations, Management

1 Introduction

The topic of desertification is an exciting subject that caused many debates, but especially polemics within international forums; the objective of this research is to contribute to bring some truths and new lightings based on our experiment and practical of the Northern margins of the Sahara desert.

It is a climatic crisis that plays the part of desertification revealing in the areas of the Sahel, (Aubreville, 1949) whereas neither official report nor desertification indexes were shown through the Maghreb countries in the 1970s.

Actually, desertification is not a new event following the dryness of the Sahel of 1970, but a phenomenon that goes back to centuries.

What is new, it is its conceptualization and its popularization under environmental phenomenon invading with the local consequences devastators, but whose world repercussions are alarming so that the world community is interested to it.

A. Marini and M. Talbi (eds.), *Desertification and Risk Analysis Using High and Medium Resolution Satellite Data,*

The myth of the desert that is moving being established, the Sahara desert is regarded as a sea of sand dunes invading the Sahel in its southern margins and consequently this projection of the desert is also imagined like a overflow of the Sahara to North, true threat on the countries of North Africa, myth largely mediatized and well maintained by politics.

On the basis of there, and from the purely scientific point of view, we will try to bring some truths from field researches on the current landscapes and to identify indicators of the paleoenvironments of these environments for better considering their evolution.

The objective is to draw the attention to the myths distorting the true debate on desertification and to confront them with the field realities to devote the effective means of the fight against desertification (Benazzouz, 2000).

The danger remains that the myth can bring the adoption of ineffective techniques of fight and cause a wasting of resources.

It is necessary to continue the efforts in order to try to delete the myths and to impose realities of the desertification.

The Algerian Atlas Mountains are drawn up like a natural barrier in northern limits of the Sahara desert.

On both sides of these mountains develops the steppe field, which includes the high plains in North and Saharan Piedmont in the South, environments where set up more or less continuous dune systems in the form of ergs.

The existence of these dunes is at the origin of the spectrum of the desertification in Algeria, phenomenon assimilated here, like a progress of the desert from South towards the North.

Which is then, the significance of these dunes and ergs sharp hanging the Atlas Mountains, on their northern slope, well beyond the limits of the Saharan desert?

How then of such sand masses often hung on the Atlas Mountains slopes could set up itself?

It is known to anybody that phenomena of desertification affect well the landscapes of the northern margins of the Algerian Sahara: "mediatized" dunes with excess with the "desert which progresses towards North" it has only one step there: the allowed idea of desertification which threatens the North of Algeria was seriously taken into account by the Algerian Authorities and was supported by the international community with the implementation of an imposing building site which is the *green Dam.*

The realization of this large green barrier in the extreme north of Sahara over a width of 20 km was to stop the progression of mobile sands of the desert invading the Sahara Atlas and the High Plains in North. Can the green dam constitute the adequate solution against desertification in Algeria?

From this question the starting assumptions:

- Does aeolian morphogenesis constitutes in fact only one form of desertification?
- Does aeolian morphogenesis contributes to desertification?

If in North Africa, it is established that one is under natural climatic conditions until the arrival of the Romans, in other deserts of the world, in particular around the edges of the deserts of the Middle-East, the processes of desertification have been ordered by anthropogenic actions and this, for at least 5,000 years, because of presence of old civilizations.

The current period, characterized by an increasingly massive presence of the man in these Atlas areas and this, since Roman colonization, falls under an evolution, which can be only anthropogenic (Birebent, 1964). Can one then distinguish the effects of a natural aridification and how to locate in time the impact of the appearance of an anthropogenic desertification in Zibans?

The current period, characterized by an increasingly massive presence of the man in these Atlas areas and this, since Roman colonization, falls under an evolution that can be only anthropogenic. Can one then distinguish the effects of a natural aridification and how to locate in time the impact of the appearance of an anthropogenic desertification in Zibans?

2 Causes of desertification in Algeria

2.1 Consequences of the historical colonization of the Saharan margins

The study of the historical colonization of the steppe and its various forms of use makes it possible to detect indices of desertification as of the Roman time through the enormous surfaces cleared and cultivated by sedentary populations (Leschi, 1943). Arab times showed a certain reconstitution of vegetation because of less pressure on the environment deduced from civilizations in matter pastoral and wandering. Since the beginning of the century, these areas have been affected by an acceleration of the degradation of the natural vegetation and soils because of the demographic pressure, agricultural mechanization, the development and the intensification of the cultures (Chabin, 1982).

2.2 Anthropic causes of the desertification in Algeria

2.2.1 Pressure of the livestock on the steppe and its Consequences

The reduction of vegetation is one of the immediate causes of desertification. However, here, the weight of the overgrazing well explains the destruction of the vegetation by the high density of the livestock in the steppe, which represented already in 1966, a sheep for 2–3 ha (Agricultural Statistics, 1968), that is to say an overload of 45%. This higher densities generate a taking away of fodder higher than the annual production of the steppe.

In addition, this average density of the livestock further increased at the end of the 1980s and at the beginning of the 1990s and is translated today, by a pastoral load of a sheep per hectare, whereas the biological standards for the steppes in the Maghreb are on the level of a sheep for 4 ha. As for manpower of the herd, they do not cease increasing, in 1970, the livestock represented 8 million heads and reached 16 million heads today (Le Houérou, 1992).

Facing this very high pastoral load, the rate of vegetable covering in the steppe hardly improved. The perennial rate of covering was 40–50% in the years 1940 to 1950 and ensured a biomass of 1,000–1,200 kg/ha. The state of current degradation of the steppe ensures only one rate of extremely weak perennial covering, hardly reaching 2–5% for a biomass from 200–300 kg/ha (Le Houérou, 1992). These results correctly translate the net retreat of the steppe, which from now on, deprived of continuous covering vegetation, will be the site of an intense wind activity.

2.2.2 The current influence of the Man on the steppe

Traditionally the practice of the pastoral activity in the steppe was subjected to the existence of a "usual wandering right" based on the obligation of rotation in the steppe. The wandering stockbreeders changed pieces of field each year according to a well-controlled organization and this, in order to ensure the reconstitution of natural vegetation. In addition to this local use of the soil, the wandering stockbreeders practiced the "Achaaba" or transhumance, practice that move twice per year approximately 2.6 million animals and 150,000 people (Agricultural Statistics, 1968). This regional complementarity ensures a connection between the high plains in north and Saharan Piedmont in the south, but remarkable aspect, implies the regular control of the pastoral load of the steppe, the control of the herd and especially a rational use of the courses.

However, the disappearance of this traditional law towards the end of the 1970s, involves a total upheaval of the steppe, where since then everyone has the right to carry out its herd where he wants and especially when he wants. The consequences result in the massive destruction of the steppe and in parallel one attends the gradual disappearance of the traditional achaaba. Today, one attends a new type of pastoralism, the "longitudinal pastoralism" which proceeds by fast displacements in the West-east direction. This modern pastoralism is the product of the thickening and the multiplication of the means of transport which allow more then undertaking, with richest to be able more quickly to bring the herd in the good grazing soils of the steppe. One imagines the catastrophic consequences well where the grass does not even have time to develop that the herd is already on the spot.

2.3 The share of the natural causes in the desertification

This fast transformation of the steppe is unfortunately accentuated here by the influence of natural factors specific to these north-Saharan margins.

2.3.1 Impact of climatic disturbances

In the high plains the prevalent direction of the winds is always influenced by the strong activity of flows from the Northwest confirmed all along the Saharan Atlas, from West to the East.

In the borders of the Moroccan frontiers, Ain Sefra (P/year: 114 mm), the resultant of the winds is from North to North-West.

On southern Piedmont of the Atlas, the prevalent direction of the winds at El Abiod Sidi Cheik station is Northern sector to North-West.

In the high plains of the Center, in Djelfa station (P/year: 335 mm) the winds blow from West and from North-West (Sary, 1973).

In the East in Hodna basin, the winds blow from West and the North-West at (Benazzouz, 1994) Bou Saada (P/year: 270 mm).

In addition, the analysis of the diurnal variations indicate that the winds of the West and the North-West act from the middle of the day at the beginning of the evening, that is to say on a ground desiccated by removing the fine particles of the soil. The morphogenesis actions of these winds are very intense in autumn on a ground practically desiccated and made pulverulent.

Consequently, the effective winds being of Northern sector in the North-West, act directly on sandy clayed soils without structure and without vegetation and this, on a large scale through the high plains.

2.3.2 The role of the topographic discontinuity of the Atlas Mountains

The Atlas Mountains do not constitute chains or barrier continues sufficiently effective to block the climatic disturbances, which are of Saharan or Mediterranean origin. Indeed, the topographic parceling out of this mountain lets appear chains in "open-type screen" or mountain with broad transverse passages (Benazzouz, 1993). These corridors develop on the topographic ensellements or structural discontinuities, sites of the Atlas cluses of the wadis which flow on Saharan Piedmont (Guiraud, 1973). Those topographic lowering of the chains, or thresholds offers the characteristic of facilitating and channeling the passage or transit of the winds of the Northern sector at the Northwest in direction of the Sahara (Benazzouz, 1994). The obviousness of these trans-atlas migrations is confirmed by the presence of many sandy plugs pasting the cluses like those of Bou Saada, (Amara and Heddouche, 1977–1978) El Haouita, (Aumassip, 1986; Estorges et al., 1969) M'Doukal (Fig. 2).

2.3.3 Effects of the endoreism in the high plains

The situation of the high plains in North of the Atlas barrier and at the same time in the axis of wind flows from the Northwest configures an open system where the winds can act with a great effectiveness (Fig. 1).

Fig. 1: Sand dunes climbing the top of Atlas Mountains

Fig. 2: The sandy plugs pasting the bed of river at M'Doukal in the Hodna basin

This wind deflation spreads by the presence of a multitude of active sebkhas occupying the center of these high plains (Fig. 1). The consequences of this total dynamics of the sebkhas appear through the construction of large aeolian accumulations on the margins of these sebkhas, in other words, there is the presence of an enormous free sand stock which can be perfectly mobilized. The origin of the dunes in the high plains finds its here explanation: it is about a local origin or produced dynamics of the sebkhas (Benazzouz, 1986).

Thus, the natural causes of the desertification in Algeria are expressed in these free sand stocks covering the high plains soils, that is to say exactly in the axis of wind flows of the North-West sector. These sands are thus in transit in the high plains and are well exposed facing the effective winds generated by these flows (Benazzouz, 1993).

The amplification of this aeolian dynamics is facilitated by an extremely advanced degradation of the steppe.

3 The role of aeolian dynamic and the processes of desertification

An imposing mountain materializes the Northern limit of the Algerian Sahara: the Saharan Atlas belt.

If one considers a progression of the desert towards North, beyond the recognized limits, it will be necessary to solve the problem of the effective crossing of this imposing obstacle and to highlight the origin and the direction of mobilization of desert sands. Which are the indices on the ground and how are registered for it flows of transfers of sands on both sides of the chain and to the center even of the chain?

3.1 The interactions relief-aeolian migrations of sand

In order to understand the interferences of the Atlas chains with reference to the sand migrations, we conducted a campaign of field investigation to identify the aeolian forms and formations through the Eastern Saharian Atlas, covering then, the Hodna basin and Zibans range. This work led to a detailed cartography of the impact on the ground of aeolian transatlasic flows. The results show two types of interference.

Mountain sets up as an obstacle with the transfers of sand, which are blocked at the northern base of the slopes, then building a lengthened erg or dune field. It is the case of the djebel Ahmar, djebel Méharga, djebel Sahbana in the south of the chott El Hodna. This barrier relief characterizes well the northern slope of the mounts of Ouled Nail where the basin of Zahrez is closed in the south by imposing dune field a 100 km length and 3–5 km wide. The on field investigation confirms a clear accumulation generalized of sands on the northern slopes with oriented dunes in Westeast to Northwest – Southeast. This situation of the dunes and the ergs plated against the Northern slopes agrees with a system of transfers of sands from the North, in the High Plains towards the South, in direction of the Sahara (Benazzouz, 1992).

The Atlas mountain presents corridors of transit and transfer of sands because to its many topographic thresholds. The most remarkable of these thresholds is located at the right of Biskra, where the Saharan Atlas shows a high lowering which facilitates the transit of mobile sands on both sides of the chains, since Hodna basin until direction of Zibans range. This aeolian trans-atlas transit is proven by many intermountain aeolian sandy ramps, which paste the tops of the mounts of Zab. Thus, the effective crossing of the Saharan Atlas by sands is solved, but with the capital difference that the mobilization of sands act from North towards the South, that is to say a supplying of the desert by sands coming from North, starting from the High Plains.

We have just highlighted a transfer of mobile sands towards the Sahara, which confirms an invasion of the Sahara desert by sands coming from the High Plains, from the North. The consequence is fundamental: there is no progress of the sand dunes towards North, it is not the desert of the Sahara which threatens north of Algeria.

3.2 The Atlas oueds: the mean source of sands

Through these oueds from Saharan margins the flow is of endoreic type and the most powerful atlas oueds reach their basic level in the sebkhas located in the high plains in North and in the sebkhas of the Sahara in the South.

However certain wadis do not cross the mountain belt of the Atlas and are lost then in the centre of intermountain depressions like those of El Outaya for the Selsou wadi. These alluvial deposits of river sediments brought closer generate vast sandy spreading on which the wind has good catch. Indeed, small ergs accompany these oueds and make up of parabolic dunes (Selsou wadi) or great barkhanes (Maiter wadi) in the Eastern Saharan Atlas. At the West and in the South of Brézina the Seggueur wadi is at the origin of the construction of one erg lengthened North-South consisted with great active dunes sharp of aklés and lengthened dunes (Callot, 1987). It acts there, as recovery by the wind of river sands or sand recycling. Thus, this wind system integrates the wadis Atlas as principal source of sand.

3.3 The place of the hydroaeolian depressions in the transatlasic transfer of sands

The Saharian Atlas is bordered of depressions parallel with the mountain belt, the Zahrez and Hodna basin in North, the corridor of the Mzi-Djedi wadi and the basin of the Low-Sahara in the South. These hydroaeolian depressions constituted of good traps for river sedimentation and are underlined by large wind accumulations. In these sites exposed to deflation, winds and floods combine themselves to ensure a permanence of sand supply of which quantities would be taken during the regional wind migrations.

Concurrently to these large cells laid out on both sides of the chains, one could highlight the existence of small hydroaeolian depressions intra-Atlas at the south-east of Hodna: bled ed-Daya and depression of El Outaya right in the North of Biskra. These depressions are formed with the favor of a localized convergence of river in a plain showing a light topographic lowering. The interest of the existence of these small hydroaeolian depressions lies in the fact that they provide sand stocks that can be mobilized and that they take turns throughout wind corridors trans-atlas. In other words, the migration of sands is not continuous in the space but translated of share the succession of these hydroaeolian depressions, an operation out of successive relays by a permanent recycling of sands.

Thus, in this vast transatlasic flux, the activity of the hydroaeolian depressions constitutes the necessary relay to the permanence of the phenomenon of the wind migrations through the Saharian Atlas. Did the significant part of aeolian dynamics in the desertification in Algeria, thus shown, how then the Man implements installations in such a significant environment and also exposed to the risks of desertification?

4 Impact of management on the Saharan environment

Of share the rigor of the climate characterized by the dryness, the arid areas were dedicated to a traditional development, which concentrated in the oases. The limited conditions of this development were compensated during time by an ancestral practice of an annual complementarily between the more rich and wetter soils of Tell Algerian mountain. This type of alternate husbandries could preserve the Saharan fragile natural ecosystems. However, during last years, the agricultural development in the Sahara, encouraged by a new legal framework (*law of the Possibility of home-ownership Land Agricultural: APFA*) knew a true revolution. The abandonment of transhumance on the one hand, and the generalization of the agricultural development on the other hand, caused a strong pressure on the fragile environment of the arid regions and strongly took part in the desertification of the Saharan margins in Algeria.

4.1 Human pressure in the oases

Demographic pressure in East Algeria is one of highest in the world with nearly 3% per year, which represents the additional arrival of a population of 700,000 per year. The Algerian population was only 4 million inhabitants at the beginning of the century, it passed from 10 million inhabitants in 1962 to 23 million in 1987 and probably to 38 million from here to the year 2000. This population is unevenly distributed, with the last census in 1987 it showed a net unbalances where 2/3 of the population concentrated in the tellien zone, in the North, that is to say 15.3 million inhabitants on just 4% of the territory. This dynamics of the population is also found in the Sahara with the nuance that the strong concentrations are located in the cities and the principal oases. The area of Zibans illustrates well the evolution of this human pressure in the oases. The population of the wilaya of Biskra was multiplied by 4 in 27 years with a total of 135,501 inhabitants in 1966, it passed to 544,978 inhabitants in 1993. It is an additional population of more than 400,000 inhabitants! This human push is marked better towards the West of Biskra where the population of Foughala was multiplied by a factor of 5.67, that is to say of 2,847 inhabitants (1966) to 16,153 inhabitants in 1993, while Doucen counted only 2,597 inhabitants in 1966, its population reaches 18,500 in 1993 thus multiplied by a factor of 7.09!

This fast increase in the population is explained by the opening of a new challenge of development in the West of Biskra, which justifies the doubling of Foughala in 6 years (4,400 inhabitants in 1987 and 8,000 in 1993).

4.2 Consequences of the overexploitation of underground waters: the case of the Low-Sahara

Among the arid regions of Algeria the Low-Sahara constitutes a rich area for the diversity and the importance of its hydraulic resources. Southern piedmont of the Atlas is irrigated by the downward wadis from Aurès mountains, while Zibans range and the oued Rhir valley, starting from sources: here there is the essential reason for the fixing of the farmers of palm trees since Antiquity period. The rational use of these water resources could preserve the natural environment during several centuries. This stability of the environment is broken following the high increase in the removal on the wadis of Aurès, the more intense exploitation of the ground water and especially the many drillings carried out in Zibans and the valley of the oued Rhir and this, in the 20 last years. The change and the degradation of the oases environment appear with the drying up of the sources and the irreversible lowering of the ground water.

The cause of the fall of the piezometric level lies in the multiplication of too brought closer drillings, which then involves the disappearance of the gushing of the sources and the wells. It is this artezianism that made the richness of the Low-Sahara.

The consequences of this degradation of the hydraulic resources are translated in Zibans by the total abandonment of many oases invaded today by sands and salts. This abundance of the hydraulic resources seems to have supported the fast increase in the population and the new conquests in the desert. This formidable interest for the Sahara concretized by Saharan agriculture often showed many disillusions.

The main part of the hydraulic resources in the Low-Sahara is fossil. A study undertaken by UNESCO between 1968 and 1972 (UNESCO, 1972) could show that the current supply of the aquifers by the rains falling on the periphery is of the same order of magnitude as the current taking away. Recently, research of A. Guendouz (1985) made it possible to specify the age of this fossil water, evaluated between

100,000 and 500,000 years, and could determine its renewal rate, considered as very weak: the current refill exists and could be evaluated on the scale of 10 years at a few centuries on the supply zones in M'Zab and in the Eastern Erg.

In the light of these results, it is obvious that any more intensive exploitation will be done with the detriment of the fossil reserves made up for the rainy periods of the Quaternary period. This recommendation does not seem to be respected and applied in the adjustments and the development projects in progress.

4.3 The recent development and the Saharan agriculture

The absence of a serious diagnosis bearing on the attacks to the natural environment following the desertification explains the failures of considerable installations.

4.3.1 Cases of El Outaya plain

It is a vast plain with 190 m of altitude, which develops just in the North of Biskra. The high content in water and in soils explains its old development. The current state of this plain shows a high rate of stranding affecting more than 60% of the surface of the plain. How could this situation develop?

Our research on the field enabled us to discover indexes of the impact on the ground of trans-atlas wind flows highlighted by surfaces of deflation, surfaces of

Fig. 3: The impact of endoreism: mobilizing of sands (yellow color) invading the irrigated field (red color) on the southern steps of Chott el Hodna Spot image, September 16th 2006

accumulation and zones of trapping of sands (Fig. 3). So, we confirmed the existence of two paleosebkhas localized to the Western termination and in the Southern edge of the plain (Fig. 3). Those paleosebkhas are located exactly in the principal direction of migration of sands through the Saharan Atlas Mountains. The development of the surfaces of sebkhas contributes to causing the reduction and even the disappearance of the transit areas of sands which will be trapped by the vegetation of these new fields very quickly stranded. More dangerous, the effect induced by this installation results in an additional sand supply in this wind corridor but also, by an insidious stranding of the oases of Zibans, in Zab Rharbi, well beyond the limits of the plain of El Outaya.

Desertification does not seem to be a concern for the Authorities because a perimeter of irrigation of 1,530 ha is envisaged in the Eastern part of the plain starting from the dam of Fontaine des Gazelles on the wadi El Hai. The site of this installation is in the surface of transit of sands and is free of sand because of its flatness, with very flattened pan and silty clayed soils. Those conditions are ideal for a maximum transit of sands because no obstacle blocks its migration. It is enough to create a new culture with the appearance of a vegetable cover and the humidification of the soil, at the time of the introduction of the perimeter so that the migration of sands is stopped and that the irrigated perimeter is transformed into fields of dunes.

In addition, the plain of El Outaya is characterized by very salted soils. The abnormally high salinity of its soils is only one recent phenomenon which appeared at the time of the colonial period and following an intensive irrigation starting from water of the wadi El Hai particularly salted and without system of drainage of salted water charged in the plain. As a proof, the soils located out of left bank of the wadi El Hai are not salted, while out of right bank, where the main part of the depression is developed, the grounds are very salted, consequences of a evil-led intensive irrigation.

This active Aeolian morphogenesis was largely facilitated by an intense development of the plain of El Outaya just like the Saharan Piedmont.

Fig. 4: Progress of dunes on the Sahara piedmont towards south in direction of Sahara desert and confirms the invading of Sahara by dunes from the North

The example of the transformation of Bled Selga plain into a sebkha with salted bottom, invaded more and more by trains of barkhanes, illustrates well the evolution of this wet bottom occupied as a preliminary by a natural vegetation at the beginning of the century. The impact of the development in this wet content facilitated the influence of the wind erosion which by deflation on the higher horizons leads to gradually export the silts and clays in the form of balls of color chestnut that one finds among mobile sands beyond the chains, on banks of the oued Djeddi (Fig. 4). The small dunes that invade the Djeddi wadi have the characteristic to be colored following the presence of two types of sand, quartz sand and silty clayed sand stocks.

The impact of the development is also translated at Saharan Piedmont by the burial of Roman ruins of the camp of Gemellae which disappears under more than 3 m of sands (Baradez, 1949).

However, it seems that it is necessary to moderate in the forms of development, particularly on Saharan Piedmont; here, the proliferation of oases in second half of the century and which cover the whole of Piedmont required important agricultural work, which is not without harmful consequences on this fragile and desert environment.

Fig. 5: Image SPOT of 21th August 2006, focused on Biskra city. The present mobiles sands are invading the Saharian piedmont of the Zibans range which appears in the lower part of the scene

Indeed, the presence of dunes with current mobile sands in the form of barkhanes constitutes the major element of the landscapes on this Piedmont and announces to some extent the doors of the desert.

These barkhanes progresses very far towards the South (Fig. 5), well beyond the position of the oases, not without having invaded and obliterated parts of them.

This phenomenon, remarkable by its width but especially by its current progression towards the Sahara, is not detectable in any cartographic document or aerial photography, only by the techniques of the remote sensing, since the spectral response of these mobile sands is infallible.

Thus, these mobile sands are clearly discriminated compared to more or less fixed older sands and the operations of optimized three-color process make it possible to be able to isolate mobile sands from other sands: the result, mobile sands appear of white color, easily recognizable, but well largely widespread on Saharan Piedmont and, most curiously, these mobile sands affect many palm plantations.

The contribution of techniques of remote sensing proves to be fundamental for studying desertification, since we could identify and locate zones of current sand mobilization in a massive way, evidence of the existence of a current active aeolian morphogenesis, codified as a form of desertification through Zibans.

These conclusions have been also corroborated by the absence of indexes of climate change for at least a century, confirmed by many work throughout the world but also in the study area, with the long series of the annual precipitations recorded at the station of Biskra, where the only variations which we could note, relate to a succession from 3 to 5–6 years from series of more or less wet years and years more or less dry.

4.3.2 The experience of Saharan agriculture in Algeria

Within the framework of the agricultural development of the territories of the South, the law on the "*Possibility of home-ownership Land Agricultural, APFA*" (Accession à la Propriété Foncière Agricole) was promulgated since 1983 and provided that in Saharan area, any ground development by a citizen gives right as property. The effects of this law made evolve the useful agricultural surface (SAU), which passed from 120,000 ha in 1983 to 420,000 ha in 1993. The evolution of the SAU hides actually the failure of the Saharan agriculture justified in fact, by economic and political reasons but could not profit from preliminary adequate treatment scientific and technical.

The assessment over 10 years is significant: the great weakness of this effective development carried out by the recipients compared to the size of the surfaces that were allotted to them. Thus, on December 31st 1993 less than 52,000 ha were emphasized, which is approximately 18% on a total surface allowed of almost 300,000 ha, through the Sahara (Adrar, Ouargla, Ghardaia, Biskra, El Oued). In the meanwhile, the mobilization of the hydraulic resource required the realization of 852 drillings of which only 70% are in exploitation on the formations of the albian series and the Continental Terminal.

However, the 10 years assessment reveals a net shift, where the resources that can be mobilized cannot meet the needs for the surfaces allowed for the development. Then, what happens of the great projects for the work and development in progress?

The assessment of execution of the land utilization in the Saharan areas engaged with the law of the APFA results in a net increase in the salinity of the soils which affects good number of the oases but also, by the appearance of the stranding in the zones, which were deprived by it. This failure confirms well the ignorance of the parameters of the desert environment, the bad conduct of the irrigation which does not envisage drainages of salted water, which leads fatally to the abandonment of these small perimeters at the end of 4 to 5 years following the appearance of salts on the surface but especially with the stranding of the unstable gardens in the wind corridors.

4.3.3 Effects of the "Green Dam" in Algeria

Launched in Algeria in 1970, the "Green Dam" project was intended to stop the relentless progression of the desert towards North Algeria and realized through a "*forest curtain*" of 1,500 km length and 20 km wide with 300 mm isohyet limit for the North and 200 mm for the South.

In 1986, the total assessment of the achievements reaches hardly 150,000 ha of surfaces treated with low rates of success from 20% to 40% on average. Since 1970 to 1980, the focus of the project was directed towards the forestation, which is carried out by the young people called of the Military Service, on the basis of annual projects of execution.

Since 1980, one can note an evolution in the approach of the green dam, which then translates into a broader idea of integrated installation limited to test areas or pilot zones.

This great work of the green dam concerns:

- Forest plantations (92,457 ha)
- Fruit-bearing plantations (5,933 ha)
- Pastoral plantations (9,080 ha)
- Work of opening of tracks (4,000 km) and installation of tracks (2,500 km)
- Work of conservation of the soils on a surface of 329,252 m^3

The general assessment of this work of forestation concerning these steppe and semi zones arid rise with:

- Surface of 1,344,273 ha all plantations
- More than 3,000,000 of ha of work of setting in defense
- 3,000,000 of ha of work of treatment of the basins slopes

This "timid" realization of the green dam expresses unquestionable difficulties, which can be explained by the non-identification and the absence of control of the manifestations of desertification on the field. The cartography of the green dam is never precise, the existing documents are limited to simple appeared of a continuous band from the Moroccan border to the Tunisian border assimilating of the biotopes as different as the mountains the piedmonts, the sebkhas, the wadis, the dunes.

It is then vital to identify in a scientific way this desert before planning to stop it and to contain it within the limits considered to be acceptable. To limit the desert starting from only one climatic parameter, in fact, here the isohyets 200 mm proves quite insufficient. The goal of the stopping was all to cover in continuous bands, one leads to a very discontinuous evolution in the space, which confirms the absence of scientific research of the surfaces favorable to the plantations devastated in addition, by the processionary worms. This major reduction in the surface really covered by the green dam eliminates the concept even of "stopping" or dam.

As for the effects of this "green dam", if they are reduced to some discontinuous surfaces of forestations facing the processes of the desertification, it seems clear that they do not have any impact on this phenomenon.

The most reached zones spread in the high plains following the evolution of the clearing of the steppe by the setting in mechanized culture, were estimated in 1986 already at 1,100,000 ha of cultivated area. It appears as well as the green dam does not induce effective protection against the desertification in Algeria.

5 Conclusion

The vastness of desert spaces can justify the bad conduct of installations, which shows the failures often repeated here.

Today, facing to the human pressure, the food challenge of country like Algeria, the error is not allowed any more, and the failure is prohibited.

Any site of future development in the Saharan areas should not constitute a surface of degradation and potential desertification, which destroys the desert environment in a irreversible way.

The knowledge of the various components of the desert environment supported by the realization of a dense network of climatic stations for measuring climatic parameters must be the priority at the moment before considering any means of fight.

In Algeria, the various forms of manifestations of the desertification are induced by anthropogenic actions too often badly led.

Bibliography

Agricultural Statistics (Statistique Agricole) 1968. *La steppe algérienne: enquête sur le Nomadisme,* n.**14**, Avril 1974 – Ministère de l'agriculture et de la réforme agraire, 385 p.

Amara A. and Heddouche Aek 1977–1978. *Existence d'un remblaiement type*, El Haouita à Bou Saada, Algérie. Libyca, Alger, t XXVI–XXVII, p 25–36.

Aubreville A. 1949. *Climat, forêts et désertification de l'Afrique tropicale.* Société d'éditions Géographiques Maritimes et Coloniales, Paris, 255 p.

Aumassip G. 1986. *Les remblaiements type El Haouita et leur signification climatique.* Trav. et doc. Orstom, 197, 11–13, Paris.

Bagnold R.A. 1941. *The physics of blown sands and desert dunes.* Methuen, London, 265 p.

Baradez J. 1949. *Fossatum Africae.* Arts et métiers Graphiques, Paris, 377 p.

Benazzouz M.T. 1986. *Recherches géomorphologiques dans les hautes plaines de l' Est algérien.* Thèse doctorat 3è cycle, Paris I, 268 p.

Benazzouz M.T. 1993. *L'endoréisme dans les hautes Plaines algériennes: Origine et impact dans les mécanismes de la désertification*, Iin colloque en l'honneur du Professeur R. COQUE. Paris 11 et 12 février 1993: Les milieux arides et semi-arides: héritages et dynamiques actuelles. Publication du Centre de Biogéographie et Ecologie, ENS Fontenay Saint-Cloud.

Benazzouz M.T. 1994. *Etude des interactions relief-migrations éoliennes de sables dans la région de M'Doukal (Algérie)*, rev. Méditerranée n. 3–4, pp 51–58.

Benazzouz M.T. 2000. *Morphogénèse holocène et actuelle dans l'Atlas Saharien algérien oriental (Algérie): conséquences sur la désertification*. Doctorat d'Etat, Université Constantine, 350 p.

Birebent J. 1964. *Aquae Romanae*. thèse université, Alger, 523 p.

Callot Y. 1987. *Géomorphologie et paléoenvironnement de l'Atlas saharien au grand erg occidental: Dynamique éolienne et paléolacs holocènes*. Thèse d'état. Paris, 412 p.

Chabin J.P. 1982. *L'homme et le milieu naturel à l'époque historique contemporaine (1850–1980) sur les confins sahariens de l'est algérien: région des Nememcha*. Thèse doctorat 3è cycle, Dijon, 520 p.

Estorges P., Aumassip G., and Dagorne A. 1969. *El Haouita,un exemple de remblaiement fini-wurmien*, Libyca, n.**17**, 53–92, Alger.

Guendouz A. 1985. *Contribution à l'étude géochimique et isotopique des nappes profondes du Sahara N.E septentrional algérien*. Thèse 3è cycle, Paris, Orsay, 243 p.

Guiraud R. 1973. *Evolution post-triasique de l'avant pays de la chaine alpine en Algérie d'après l'étude du Hodna et des régions voisines*. Thèse doctorat es-sciences, Nice, 270 p.

Hans-Jurgen Bolle 1996. *Global climate change and desertification*. In 28th Conf. Intern. of Geography, 4–10 August, Den Haag.

Le Houerou H.N. 1968. *La désertisation du Sahara septentrional et des steppes limitrophes*. Ann. alg. de géog., Alger n.**6**, 5–30.

Le Houerou H.N. 1992. *Recherches biogéographiques sur les steppes du Nord de l'Afrique*. Thèse doctorat d'Etat, université Paul Valéry, Montpellier, 1992.

Leschi L. 1943. *Le «Centenarium» d'Aqua Viva près de M'Doukal (Commune mixte de Barika)* Revue Africaine, n. 394–395, pp 5–22.

Sary M. 1976. *Géographie physique d'une Haute Plaine steppique algérienne: le Hodna*. Thèse 3è cycle, Strasbourg, 284 p.

Shaw B.S. 1981. Climate, environment and history: the case of Roman North Africa, in Climate and history studies in past climates and their impact on Man, T.M.L. Wigley, M.J. Ingram, G. Farmer eds., Cambridge University Press, New York, pp 379–403.

Trayssac J. 1981. *Etude géomorphologique du bassin-versant de l'oued Djelfa-Melah, versant Nord des monts des Ouled Nail (Algérie)*. Thèse 3è cycle, Poitiers, 211 p.

Trayssac J. 1983. *Migrations de sables éoliens depuis l'Holocène dans les Monts des Ouled Nail (Algérie). Paléoécologie des régions sahariennes*, Actes du Colloque international, Béni Abbès, 20–30 octobre 1983, CNEH, Alger, p 28–39.

UNESCO 1972. *Projet «ERESS» Etude des ressources en eau du Sahara septentrional*. Rapport final, Paris.

Williams G.E. 1970. *Piemont sedimentation and late quaternary chronology in the Biskra region of the Northern Sahara*, Z.f. geom. N.F., n.**10**, 40–63, Berlin.

CLIMATE CHANGE AND THE NEED FOR RENEWABLE ENERGY SOURCES

AMINA MALAKI, ABDELILAH EL GHANNOUCHI, AND MOHAMED EL WARTITI
Département des Sciences de la Terre, Faculté des Sciences UFR R.M.R.E.E. Rabat
Université Mohammed V –Agdal-Rabat
Amina_ok@yahoo.fr

Abstract

Scientists have drawn the bell from alarm for more than 20 years and warned us: the climatic extent and the speed of the change of human origin are without precedent and make heavy threats on everything that depends on the climatic conditions. The quality of life of many populations, biological diversity, harvests, glaciers and ice-barriers, the sea level, the frequency and intensity of the weather catastrophes (storms, floods…). How to produce then the quantities of energy necessary to satisfy world consumption without deteriorating in an irreversible way the environment? Renewable energies – wood, wind, sun, small hydraulic – can play an important part in the durable development.

Keywords: Climatic change, durable development, renewable energies

1 Introduction

Greenhouse gases are accumulating in Earth's atmosphere as a result of human activities, causing surface air temperatures and subsurface ocean temperatures to rise. Temperatures are, in fact, rising (Fig. 1).

The changes observed over the last several decades are most likely due to human activities, but we cannot rule out that a significant part of these changes is also a reflection of natural variability. Human induced warming and associated sea level rises are expected to continue through the 21st century. Computer model simulations and basic physical reasoning suggest secondary effects. These include increases in rainfall rates and increased susceptibility of semi arid regions to drought. The impacts of these changes will be critically dependent on the magnitude of the warming and the rate with which it occurs.

Of the greenhouse gases that are directly influenced by human activity, the most important are carbon dioxide, methane, ozone, nitrous oxide, and chlorofluoro-carbons (CFCs). Aerosols released by human activities are also capable of influencing climate.

A. Marini and M. Talbi (eds.), *Desertification and Risk Analysis Using High and Medium Resolution Satellite Data*,

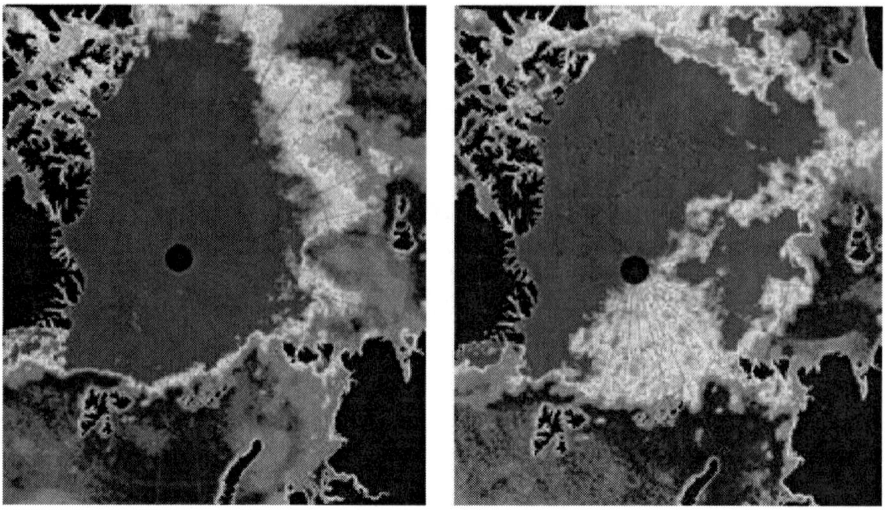

Fig. 1: The Arctic ice, seen by EOS Aqua In August 2005 (left) and August 2006 (right)

2 Dangerous climate impacts and the Kyoto protocol

Climate change scenarios for Africa, based on results from several general circulation models using data collated by the Intergovernmental Panel on Climate Change (IPCC) Data Distribution Center (DDC), indicate future warming across Africa ranging from 0.2°C per decade (low scenario) to more than 0.5°C per decade (high scenario). This warming is greatest over the interior of semi-arid margins of the Sahara and central southern Africa (Cavard et al., 2000). Africa is highly vulnerable to the various manifestations of climate change. Six situations that are particularly important are:

- Water resources, especially in international shared basins where there is a potential for conflict and a need for regional coordination in water management
- Food security at risk from declines in agricultural production and uncertain climate
- Natural resources productivity at risk and biodiversity that may be irreversibly lost
- Vector- and water-borne diseases, especially in areas with inadequate health infrastructure
- Coastal zones vulnerable to sea-level rise, particularly roads, bridges, buildings, and other infrastructure that are exposed to flooding and other extreme events
- Exacerbation of desertification by changes in rainfall and intensified land use

Climate change and desertification remain inextricably linked through feedbacks between land degradation and precipitation. Climate change may exacerbate desertification through alteration of spatial and temporal patterns in temperature, rainfall, solar insolation, and winds. Conversely, desertification aggravates carbon dioxide (CO_2)-induced climate change through the release of CO_2 from cleared and dead vegetation and reduction of the carbon sequestration potential of desertified land. Although the relative importance of climatic and anthropogenic factors in causing desertification remains unresolved, evidence shows that certain arid, semi-arid, and dry subhumid areas have experienced declines in rainfall, resulting in decreases in soil fertility and agricultural, livestock, forest, and rangeland production. Ultimately, these adverse impacts lead to socioeconomic and political instability. Potential increases in the frequency and severity of drought are likely to exacerbate desertification (Sokona et al., 1999).

Water resources are inextricably linked with climate, so the prospect of global climate change has serious implications for water resources and regional development (Polidano et al., 2000).

In the Maghreb region the most striking effect of climate change will be the reduction of rainfalls. This is a direct threat to Morocco's lifeline. If the declining trend continues, Morocco is going to loose its agriculture. This will constitute tremendous economic, social and cultural damages. The Sahara will move across the Atlas (Fig. 2). How fast this process will progress in future does depend on the development of worldwide CO_2 emissions, i.e. is subject to human influence (Gerere, 1993).

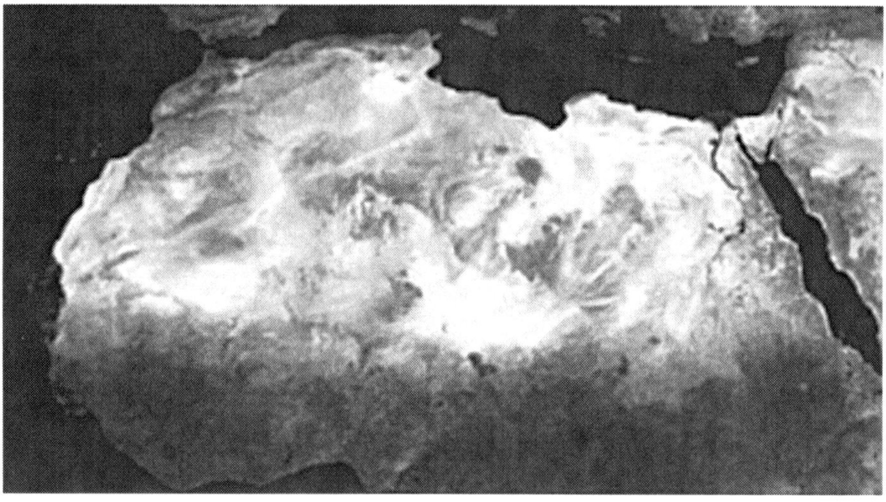

Fig. 2: Satellite image of the Sahara

The continued extension of worldwide CO_2 emissions is a direct attack against Morocco, contributing to make it an inhabitable country. Morocco has to take measures – either dealing with the damages or preventing them. In Morocco, the northward displacement of the Azores high-pressure cell is a subject of study because of its association with a drought cycle that is related to the dipole between the positioning of the Azores high-pressure cell and the Iceland low-pressure cell. These severe droughts seem to manifest themselves in Morocco in periodicities varying between 2 and 13 years.

A global strategy on climate change has been agreed under the 1992 United Nations Climate Change Convention and its 1997 Kyoto Protocol. This international legal regime promotes financial and technical cooperation to enable all countries to adopt more climate friendly policies and technologies. It also sets targets and timetables for emission reductions by developed countries. Specifically, the Protocol requires 39 developed countries to reduce their greenhouse gas (GHG) emissions by an average of 5.2% compared to 1990 levels. These annual emissions reductions must be reached by 2008–2012, referred to as the first commitment period.

The developed countries with emission reduction targets are called Annex I countries (industrialized countries), whereas those without targets are the non Annex I countries (developing countries) (Polidano et al., 2000).

The Kyoto Protocol allows developed countries to reach their targets in different ways. They have to implement domestic Policies And Measures (PAM) and they are allowed to supplement these PAM with projects abroad and with market instruments, also called flexibility mechanisms as the Clean Development mechanism (CDM).

The CDM, defined in Article 12, allows Annex I Parties to implement projects that reduce greenhouse gas emissions in non Annex I Parties and has the double goal of assisting non Annex I Parties to achieve sustainable development and contribute to the ultimate objective of the Convention, i.e. to stabilize greenhouse gas emissions at levels not dangerous for human development (Austin et al., 2000).

To contribute to the effort of the international Community to reduce the climate change impact, Morocco signed the UNFCCC in 1992 and ratified it in 1995. Morocco also ratified the Kyoto Protocol in January 2002. A portfolio of four projects has been developed for the reduction of GHG emission, two of which are in the area of renewable energy (Matte, 2005).

The table below presents a summary of this portfolio (Figs. 3 and 4).

Project title	Annual reduction of CO2e (t)
Essaouira windpower project	156.026
Tétouan Wind Farm Project for Lafarge Cement Plant	28.651
Photovoltaic kits to light rural households in Morocco	38.636

Fig. 3: Projects in Morocco registered by the CDM Executive Board as at 8 May 2006. CO_2e= carbon dioxide equivalent (*UNFCCC* website)

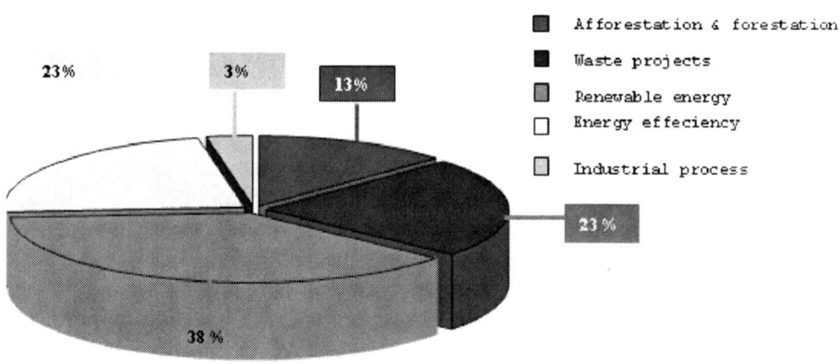

Fig. 4: Preliminary CDM projects portfolio

Half of the Moroccan portfolio concerns wind energy projects, which are described below in detail.

– 10 MW wind farm for the LAFARGE-MAROC maroc cement factory at Tétouan.

LAFARGE-MAROC, the largest cement company in Morocco, has a project to build a 10MW wind farm to produce power for its new cement factory at Tétouan in the Rif region. This project should reduce CO2 emissions by approximately 33,000 t per year, that is a total of around 8 million tons during the 25-year life span of the project (Matte, 2005).

– 60 MW wind farm at ESSAOUIRA

The proposal of the Moroccan National Electricity Board (ONE) is to construct a 60 MW wind farm near the town of Essaouira to supply electricity to the national grid. Preliminary studies for the project indicate an annual electricity production of 210 GWh and a reduction of CO2 emissions of approximately 4.9 million tons during the 25-year life span of the project (Matte, 2005).

Extremely good wind and solar radiation resources, their supplementary nature to the European ones, and the close neighborhood to Europe put Morocco into a strong position.

Southern Morocco has large regions with excellent solar and wind conditions (Knies et al., 2002):

1. The wind potential: In the Southern parts of Morocco, in trade wind regions with a yield between 3,000 and over 4,000 h/a full load, an electric power of over 1,000 TWh/a could be generated.
2. The solar potential: Typical values for the annual solar direct normal irradiation are 2.5 MWh/m and more or 2,200 and more full load hours per year. Using thermal storage the full load hours for solar power plants can be extended to over 5,000 h/a, depending on the collector field size.

3 Renewable resources of the south and east Mediterranean region

The South and East Mediterranean region has a wide potential for development of renewable energy because of its large population and vast solar and wind resources (Fig. 5). Moreover, in order to supply the energy and electricity needed for the social and economic development of the countries of the region, while preserving the environment, it is essential to promote advanced energy technologies, in particular, renewable energy. The important roles that renewable energy could play in contributing towards sustainable development and poverty alleviation were also confirmed at the Johannesburg World Summit on Sustainable Development.

The South and East Mediterranean countries are endowed with high potential of renewable energy sources and these resources are underdeveloped because of many existing barriers. The projects also showed that CDM could play a positive and important role in developing the renewable energy market in the South and East Mediterranean region, thereby contributing to the sustainable development of the region. But despite high renewable energy resources, it confirmed the lack of adapted institutional frameworks for the large-scale integration of wind power and PV in the Southern and Eastern Mediterranean countries (SEMC), and the need for

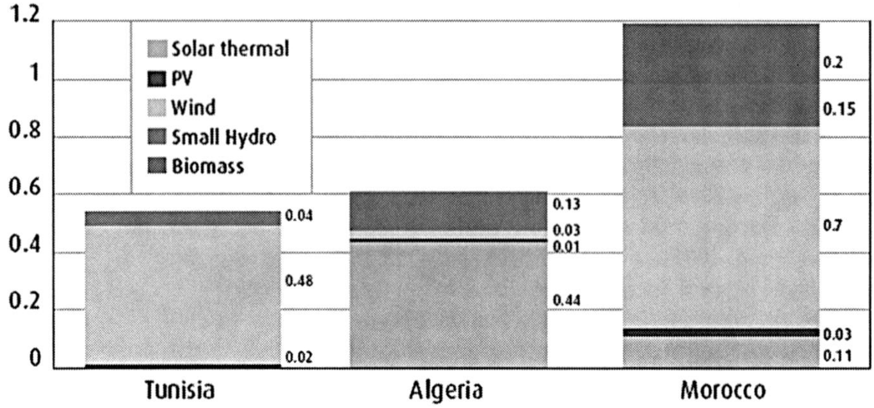

Fig. 5: Renewable energy sources (excluding large hydro)-based power generation in 'MODRES 2010 SCENARIO' (in TWh) (by source and country)

specific financing schemes, such as the CDM, to allow the renewable energy market to significantly develop in the Mediterranean region (Austin et al., 2000).

We recognize the importance of renewable energy for sustainable development, diversification of energy supply, and preservation of the environment.

Renewable energy resources can contribute to sustainable development, but important barriers impede the realization of their potential (Sokona et al., 1999).

There are four main obstacles to incentives and institutional approaches: regulations are lacking, there is no dedicated agency, RE are low priorities for national development programs dedicated to raising awareness and to ensuring sustainable demand in RE technologies and services, and taxation do not provide attractive market conditions. Finally, the added value of RE in employment and in attracting new investments is not taken into account.

The mechanism developed by public authorities to promote renewable energy will be highly favorable to the achievement of the objectives through relevant reforms:

- Operational regulations, which more fully integrate energy and environment
- Strategies to pool private operators in favor of more available quality equipment and services, of adapted financial instruments and mechanisms to ensure proper alignment of stakeholder initiatives and genuine market sustainability
- Support and monitoring tools, too often neglected in renewable energy
- policies

Although renewable energy resources are now being utilized more than ever on a global scale, there is no doubt that their contribution to the energy economy can still be greatly increased. Recently international support for developing these relatively new sources of energy has been driven by their benefits as assessed by reduced environmental impact, particularly reduced greenhouse gas emissions.

4 Conclusion

Renewable energy technologies have great capacity to contribute to other aspects of sustainable development. The extent of investment into renewable energy induced by the Kyoto mechanisms will depend on whether the rules and guidelines that are to be developed in the coming years will explicitly support renewable energy. The Kyoto mechanisms could be instrumental in leading to significant investment into these resources if rules are defined appropriately.

Achieving solutions to environmental problems that we face today requires long-term potential actions for sustainable development. In this regard, renewable energy resources appear to be the one of the most efficient and effective solutions. Therefore, clean, domestic and renewable energy is commonly accepted as key for future life for Morocco.

Remarkable progress has been made in the commercialization of renewable energy options. Wind power, for example, is a success story and may soon be the most economic supply alternative. Solar energy is coming of age and unit costs are coming down at an impressive rate. In the long run, renewable energy will be the major contributor to the world's energy system. The industrialized countries of the North have most of the technologies and the financial resources for utilizing renewable energy technologies, while many developing countries have great potential for renewable resources. Therefore, technology transfer to developing countries is needed, and the Kyoto mechanisms could play an important role.

Bibliography

Austin D., Faeth P., 2000: *Financing Sustainable Development with the Clean Development Mechanism*, World Resources Institute, Washington, DC.

Cavard D., Cornut P., Menanteau P., 2000: *Pays en développement et la prévention du risque climatiques: Quelles perspectives pour le mécanisme de développement propre?* in Revue de l'Energie, numéro 520, p. 464–475.

Gerere, 1993: *Energie et environnement au Maroc, étude pour la Japanese International Cooperation Agency (JICA),* Rabat.

Knies G., Bennouna A., 2002: *Potential of Moroccan Renewable Energies for the Defence Against Climate Change*. Forum International sur les Energies Renouvelables, Tétouan, 8–10 mai 2002.

Matte, 2005: *Portefeuille de Projets MDP, Ministére de l'Aménagement du Territoire de l'Eau et de l'Environnement,* Projet RC MDP Maroc, 2005.

Polidano C., Jotzo F., Heyhoe E., Jakeman G., Woffenden K., Fisher B.S., 2000: *The Kyoto Protocol and Developing Countries: Impacts and Implications for Mechanism Design*, ABARE research report 2000.4, Canberra, June 2000.

Sokona Y., Humphreys S., Thomas J.P., 1999: *The Clean Development Mechanism: What Prospects for Africa?* Environnement et développement du Tiers Monde (ENDA-TM), Dakar, Sénégal, 25pp.

CHANGES IN THE RESOURCES MANAGEMENT AND THEIR ENVIRONMENTAL CONSEQUENCES IN A SAHARAN ARID ENVIRONMENT

THE CASE OF THE JEFFARA PLAIN, THE NEFZAOUA OASES AND THE SEHIB AREA IN SOUTHERN TUNISIA

MOHAMED TALBI, NAJET BEN-MANSOUR, KHALED TALBI, AND NABIL GASMI
Institut des Régions Arides. Medenine. 4119. Tunisie
<Mohamed.Talbi@ira.rnrt.tn>

Abstract

Since the earliest antiquity, natural resources in southern Tunisia have been used within the scope of a collective organization and of a balanced and sustainable management. However, from the beginning of the 20th century, the local populations way of life underwent changes that entailed the occurrence and development of a new way in using resources (especially regarding water and land), based on individual initiative and often regardless of the social regulations in force.

The present study highlights from one hand environmental changes from case studies regarding the *Jeffara* (south-east), and the *Nefzaoua* (south-west), Tunisian regions that have been suffering unprecedented degradation of natural resources, due to various pressures.

From the other hand, a third case study, regarding the *Sehib* area (southern low plains) is an example of an environment that has kept its balance, thanks to the population well thought-out and sustainable management of its natural resources. This work used multi-source data, like aerial photographs, satellite images from different dates, auxiliary data and field work. We have studied the management of natural resources in relation to the potential of the environment on the one hand, and to the needs of the concerned populations on the other hand, through a monitoring of the spatial dynamics of land use and land cover in southern Tunisia.

The aim is to define the areas affected by important dynamics, to draw the attention to the need of a strategic, integrated and joint approach of the natural resources management. Such management should particularly rely on the *protection* and *restoration* of *local knowledge, as well as* the correction of the *mistaken assessments* of the "*official science*".

Keywords: Desertification observation, natural resources degradation, remote sensing and information systems, social transformations in arid environment, local knowledge, Tunisia, Maghreb, Mediterranean

A. Marini and M. Talbi (eds.), *Desertification and Risk Analysis Using High and Medium Resolution Satellite Data,*

1 Introduction

The study areas are located in southern Tunisia (Figs. 1 and 2); and are affected by intense desertification with dramatic consequences on the physical and human environment. In Tunisia, as soon as 1972, on the occasion of the *Gabes Symposium*

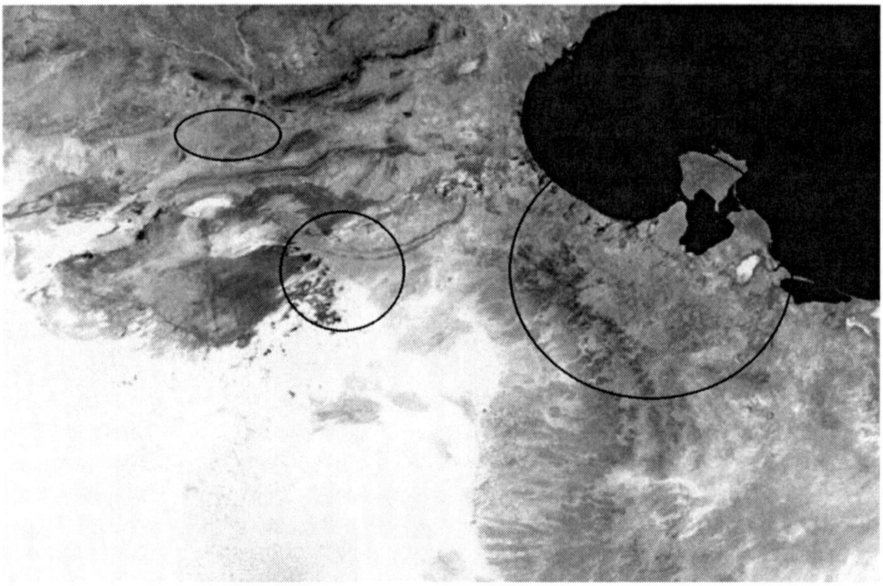

Fig. 1: Localization of the study areas on a RESURS-01 satellite image

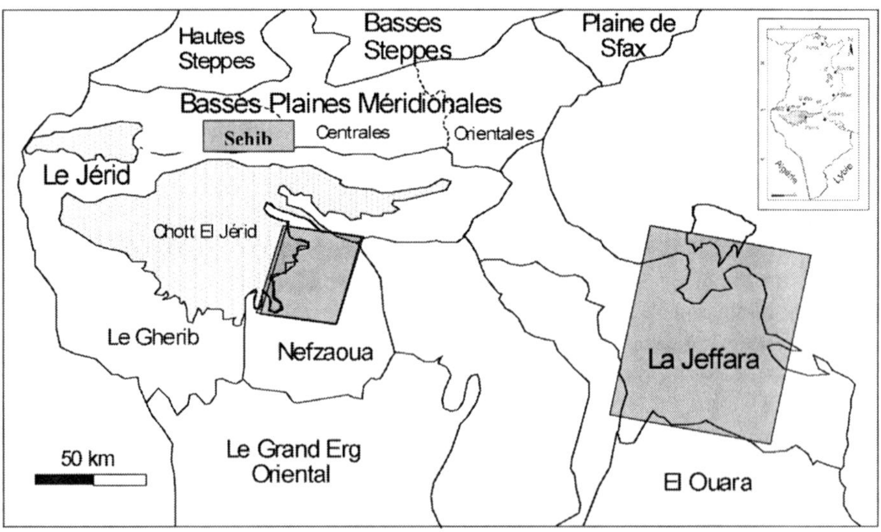

Fig. 2: Natural regions in southern Tunisia. After H.N. Le Héouerou

on land degradation, studies dealing with this phenomenon gave alarmist reports. Later in 1977, H.A. Dregne in its *world map of desertification*, considered that Tunisian arid area was subject to severe desertification. In that map, the only African area classified as affected by very serious desertification was situated in southern Tunisia. Historical sources agreed that this region has always been an area of settlement and a crossing point between the sea and the African continent (caravan trading), and between the North, the far South and the Middle East (Mzabi, 1988). Nevertheless, the area kept a relative balance between resources and management.

So what caused that such important degradation phase in the southern Tunisian environment, which was unknown up to the present day?

Is it a crisis due to a long-term drought, as was believed at the beginning of the 1970s? Or is it a phenomenon mainly induced by human actions?

At first sight, the features of this Tunisian region seem to predispose it to land degradation and desertification, e.g. climate (long dry season, considerable water deficit), lands (fragile inherited soils), and flora (poor steppe vegetation).

Nevertheless, and according to most research works (Rapp et al., 1976; Khatelli, 1981; Floret and Pontanier, 1982; J'Girim, 2000), climatic conditions did not significantly change during the last two millenniums. Consequently, the main causes of degradation seem to be attributed to human actions, mainly induced by changes in the local population's ways of living.

Indeed, many studies already dealt with on the evolution of the population's way of life and its consequences on the uses of resources (Louis, 1979). Research dealing with this subject, especially by the Institute of Arid Regions (Abaab, 1981, 1984; Omrani, 1982; Bouhaouach, 1983; Talbi, 1983, 1993, 2000; Nasr, 1993; Meat, 1996) determines three key phases, i.e.: a *nomadic phase*, or balance phase, from the 11th century to the end of the 19th century; a *semi-nomadic phase* (the beginning of the imbalance) that began with the French colonization; a *sedentary phase* – an overall disruption with the rules that linked men to their environment.

Hereafter, we will sum up the important events of these various periods and derive their impacts on the environmental dynamics of the studied area.

2 From a collective management to individual use of resources

We will evoke the general features of men's behavior towards their natural environment, from the nomadic to the colonial phase.

2.1 Nomadic phase

During the period dating from the *Hilalian invasion* (11th century) to the end of the 19th century, nomadic tribes used to live in rangelands located between the natural regions of *Matmatas* and the coast, for the populations of the *Jeffara* area, and in the regions situated between the *Dahar*, the *Erg* and the *Gherib*, for the nomads of the *Nefzaoua* area (Fig. 2).

The local tribes used to live from the extensive breeding of sheep, goats and camels; they followed the rains and moved all over the territory. They had *no fixed territory,* as remarked by Bouhaouache (1983).

On the whole, the regional economy was agro-pastoral, if one considers that the pastoral activity of nomads prevailed, but that its survival was conditioned by the farming activities of the sedentary populations in the *Matmata* Mountains and the *Kebili* peninsula, who supplied nomads with part of the necessary food.

Within a tribe (such as *M'Razigues* in *Nefzaoua*) or a confederation of tribes (such as *Ouerghemma*), the relationships between the various members were governed by *clannishness*, which consisted in the adherence of every member of the tribe to the choices and decisions regarding the lot of the whole social group (Abaab, 1984).

The consequences on the environment were namely that there was no settling process, except for the *Kebili* peninsula and the areas of *Matmatas* and *Jerba* Island.

Consequently, the *Jeffara* and *Nefzaoua* regions were places of nomadism for tribes whose self-subsistence allowed them to be partly autonomous in relation to the outside world.

Abaab (1984) underlines that « *such a system was autarchic and precarious, but was remarkably well adapted to the requirements and constraints of arid environments, because tribes were constantly moving and there were no legal or political obstacles (frontiers) to the movements of men and their herds* ».

That is why, in such a system, lands were used and food was looked for at the scale of generations. In places where there were few rains, dry farming (cereals) or breeding were undertaken in places of higher rainfall. This integration of rangelands and crops into land ecology succeeded in being maintained for centuries.

But this system was to be disturbed at the end of the 19th century, with the beginning of the *French colonization* in southern Tunisia.

2.2 Semi-nomadic or colonial period

The colonial authorities declared this part of the Tunisian territory to be a "*military area*". This is why the protectorate army had to control the movements of tribes, in order to keep them under control. Such a policy was to trigger off a process of destruction of production systems and nomadic social structures. This is particularly due to the implementation of programs of systematic populations settlement, to the creation of new systems of farming production and to the introduction of money exchanges.

The settling process was intended to allow the authorities to improve their control of local populations. With this aim, the colonial army began to privatize lands (which were indivisible until that time), and to break up the systems of community appropriation of lands and resources.

In the meantime, the colonial authorities had created many watering places, villages, towns and farm markets in the *Jeffara* and *Nefzaoua* regions.

These first built-up areas, with a basic infrastructure (usually a well) and a network of passable tracks, became a magnet for the nomadic populations.

According to Louis (1979), this colonial period entailed a considerable change in the distribution of the populations and in the economic activities of the *Jeffara* and *Nefzaoua* regions. For Abaab (1984), this policy « *aimed at integrating the populations of the region into the area of commercial domination of the capitalist economy, as well as at opening southern Tunisia to colonial goods (sugar, tea, coffee, spices) in view of progressively transforming the consumer habits of the local populations ».*

As the environment could not feed the whole population, the imbalance forced people to migrate, according to the seasons (individually or in groups), in order to look for food, mainly to regions producing olives, dates and cereals. This is the cycle of "*HTAYA*", a traditional migration in the *Jeffara* region (Fig. 3).

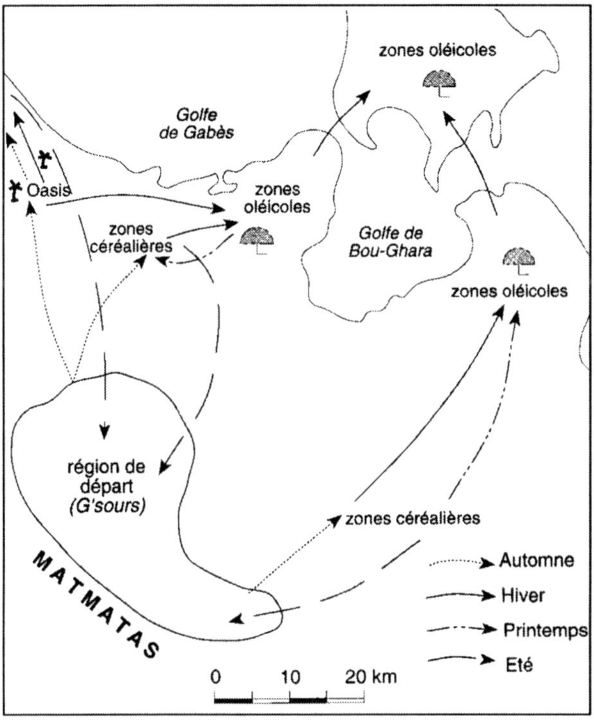

Fig. 3: The Htaya: traditional migration cycle

Autumn: group migration to oases and cereal-producing areas
Winter: group migration to olive-producing regions
Spring: group migration to cereal-producing areas (harvesting)
Summer: return to original region. Preparation of next campaign

In conclusion, this semi-nomadic or colonial phase was characterized by the delimitation of collective lands, which were indivisible until that time; the difficulty to practice transhumance and nomadism; a disruption in the nomadic way of life, by depriving it from the necessary natural resources, thus compelling nomadic populations to settle down. Consequently, the livestock moved less and grazed in limited areas. The distribution of collective lands transformed the landscape (creation of oases, irrigated crops, dry farming, and cereals cultivation).

Therefore, it may be considered that during this period balance was precarious, and became severely and sometimes irreparably broken down by men, who clashed with their environment during the sedentary (or current) period.

The following chapter bears on the analysis of the transformations that affected the *Jeffara* and *Nefzaoua* regions, and of their consequences on environmental dynamics, on the state of natural resources and on the quality of life of the concerned populations.

3 Analysis of the environmental dynamics and its consequences on the environment

3.1 Tools and method

We will deal with notions of compilation and analysis of reference data to derive thematic information. In our case, three sources provide reference data: the existing literature, unpublished documents (or material collected within the scope of the study), and then the culture and knowledge of the expert (or research team) who is both interested in research and novel knowledge and has the skill to ask differently questions whose answers are considered to be already known.

However, it is essential to prevent the expert knowledge from hindering the understanding of phenomena.

Regarding our study, there are many bibliographic data, but they are sectional and highly inhomogeneous.

As far as maps are concerned, we use Survey maps (1/200,000 scale) from 1905, which are rightly considered as the first reference documents.

For comparison needs, we used the "new" 1/200,000 scale maps, drawn in 1977, using aerial photographs taken in 1971. We consider that these last documents witness the situation, just before the big changes.

Regarding the *Jeffara* region, we use two *LANDSAT* satellite images, the first one dating back to the beginning of the 1970s (MSS image of November 8th; 1972) and the second one is a *TM* image of January 27th, 1985, as well as a recent LANDSAT-7 TM image taken in January 2000.

We also used aerial photos at different scales (from 1/85,000 to 1/10,000), taken in 1963, 1971, 1974 and 1985 aerial missions.

Regarding the *Nefzaoua* area, for the period when big changes began, we use three *Landsat TM* satellite images from 1985, 1996 and 2000. We selected our satellite images of *May* and *June*, best period of the year to monitor irrigated areas. For the same study, we also used aerial photos from various missions: 1971, 1985, and 1998.

3.2 Work performed

Thematic information was derived from collected, compiled and synthesized multi-source data. The various stages were the following: collecting available bibliographic documents and maps, field survey and inquiries with people who witnessed the changes recorded; selecting aerial missions, getting the available aerial missions (1967, 1971, 1985, 1998) and outlining photo-interpretation; selecting satellite images available from different providers (*Landsat, SPOT, IRS, Resurs*, …) and obtaining those that are likely to bring answers; pre-processing images (radiometric and geometric corrections, enhancing contrast,…) and ground control points (*GPS*, …);

testing various algorithms and choosing processes adapted to the theme and region studied; deriving information, mapping, and validating results.

At this stage, synthetic maps should combine several space variables on the same background.

The links of causality or dependence that exist between several data are expressed on these variables, as well as their possible capacity to determine as a whole other phenomena or combinations.

Our aim in this study is to classify the *Jeffara* and *Nefzaoua* regions into homogeneous areas; i.e. areas with minimum differences or conversely maximum similarities. Such homogeneous areas are not presented statically or descriptively, but to be used efficiently for *predictive mapping. A predictive or prospective approach* means an approach aimed at forecasting the future of a place (or its dynamics) and at drawing the attention to the importance of hazards likely to affect the environment.

3.3 Results and discussions

This part is devoted to analyzing of the main trends of the environmental dynamics in the *Jeffara* and the *Nefzaoua* regions; monitoring the evolution of phenomena and "*detecting changes*" occurred between different images from different dates (month, season, year or decade).

With the availability of space imaging and the quick development of utilitarian data processing, it is now possible to manage and compare lots of data in a rapid, reliable and "neutral" way, which was not feasible until recently.

4 Landscape dynamics in the *Jeffara* region

When Tunisia became independent in 1956, the south of this country faced a difficulty economic situation (Mzabi, 1988). Except the farming sector that employed more than half the working population, there was quite no other activity likely to provide a source of "salaried" income for several tens of thousands unemployed people. This time was marked by two periods:

4.1 From 1956 to 1970

During this period, the environment faced a slow inexorable degradation, but kept a certain balance. The reasons of this evolution are extremely important and should be mentioned:

Between 1956 and 1962, it appears that the quite general abandonment of lands and pastoral activities was induced by the attraction of recently created urbanized areas (watering places, possibility of earning wages…) and the migration of thousands of working people from the *Jeffara* and *Nefzaoua* regions to other areas of the country and abroad.

Between 1963 and September 1969, Tunisia experienced a "*collectivist*" policy in development which was wrongly implemented and finally ruined the country, inducing a public mood of distrust that took farmers away from agriculture and pastoralism.

The social situation was dramatic: peasants were ruined, exodus to towns intensified and state works to fight against unemployment were unable to meet all job applications. There were various consequences on the physical environment, from the abandonment of plantations that were likely to be collectivized, to the lesser pressure on the environment due to the regression of the stocking rate (half of the livestock had been slaughtered or sold at derisory prices between 1966 and 1969 (Talbi, 1993).

But in 1969, the abandonment of the "*collectivist*" experiment, triggered off two activities that are important causes of desertification: the comeback of breeding, and the unprecedented development of land clearing using heavy machines. This was possible especially due to financial transfers from thousands of workers originating from the region who lived abroad.

4.2 From 1970 to date

A short waiting time after the abandonment of the co-operative way and the restoration of a free-market economy, the money coming from abroad played a considerable part in the sudden change that affected the *Jeffara* region as soon as the beginning of the 1970s (Nasr, 1993). This financial power mainly focused on four sectors: settling process; allocation and cultivation of new lands; comeback of breeding and generalized mechanization.

Of course, these four sectors evidence the revival of the regional economic and agricultural dynamics; however, they are the key factors of the large environmental degradation in the main types of landscape of the *Jeffara* region (Figs. 4 and 5).

4.3 Settling process

The settling process was promoted by the colonial authorities and then by the independent governments, but it became important only with the financial support from the regional workers who had emigrated abroad. In fact, « *the first reflex* » of rural migrant in *Europe* was to build a permanent house on their lands.

This resulted in a rural scattered but dense settlement, which now characterizes the *Jeffara* region in comparison with the other arid areas of the country.

As consequence, this rural permanent settlement directly entailed the maintenance of a constant human and animal presence on the whole territory studied, which was expressed by a considerable continuous pressure on the environment. Indeed, this type of « *rural settlement* » is more detrimental to the environment than urbanization.

4.4 Allocation and cultivation of new lands

This trend considerably increased at the beginning of the 1970s, mainly because of the beginning of the process of privatizing collective lands, and of the increased purchasing power of migrant workers abroad.

Indeed, the « *second reflex* » of emigrants, after building a permanent house was to increase their real estate, by purchasing or renting new lands, in order to plant olive trees and to cultivate cereals. This type of cultivation induces a mobile surface

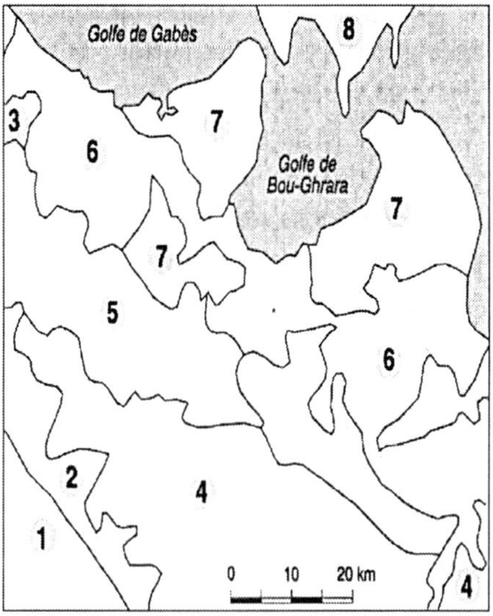

Fig. 4: Main types of landscape in the Jeffara region **1**: mountains, **2**: piedmont, **3**: oases, **4**: sandy plains, **5**: crusted Medenine plateau, **6**: salty gypseous environment, **7**: olive groves, **8**: Jerba Island.

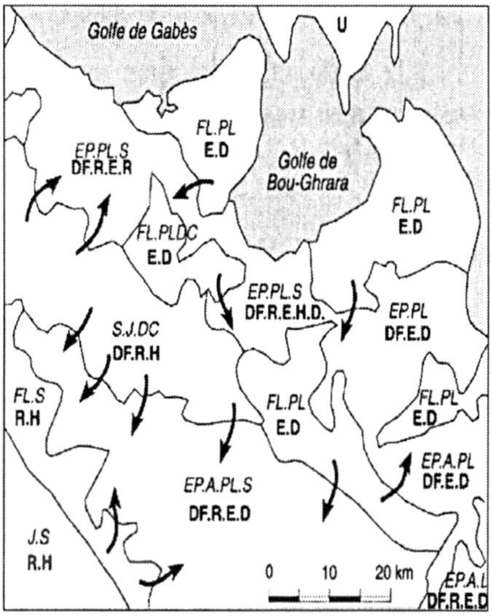

Fig. 5: Causes of degradation and their environmental consequences
1-Causes of degradation: **EP** – allocation of collective lands; **A** – extension of tree cultivation; **PL** – use of unsuitable tools; **FL** – frequency of ploughing (dry farming), **S** – overgrazing – poor maintenance of structure works; **DC** – destruction of calcareous crust; **U** – urbanization.
2-Consequences on the environment: **DF** – clearing of rangelands; **R** – regression of natural vegetation cover, **E** – wind erosion; **H** – water erosion, **D** – formation of mobile dunes.
2-Conséquences sur le milieu: **DF** – défrichement des parcours, **R** – régression du couvert végétal naturel, **E** – érosion éolienne, **H** – érosion hydrique, **D** – formation des dunes mobiles.

soil and destroys the vegetation cover while depriving it from its humus reserves. This is why, in the most wind-exposed places, sand accumulation rapidly forms dunes from this eroded material.

4.4.1 Frantic mechanisation

At the beginning of the 1970s, with the abandonment of the "*collectivist*" experiment and the comeback of agriculture, the use of tractors had become widespread.

Swing ploughs and ploughshare ploughs were substituted by *poly-discs* that induced considerable water and wind erosion (with a soil loss ranging from 50 to 250 t per ha per year, according to a study by the Institute of Arid Regions).

This rapidly led to a desertified landscape. Khatelli (1981) and Novikoff (1983), evidenced that the daily working capacity of a tractor is forty times higher than an animal.

As a consequence, a hundred tractors may cause irremediable damage to 300,000 hectares within a single plough campaign in case of a rainy year.

Since 1972, the Institute of Arid Regions has been leading studies regarding the inventory and monitoring of desertification, particularly through remote sensing. These studies actually show the devastating consequences of farming machines; especially poly-discs plough (Talbi, 1993).

Therefore, mechanization entailed considerable adverse results because of the use of tools non-adapted to the climatic conditions and the type of soils. This is why, within few decades, damages were more important than those suffered by the region for whole centuries.

4.5 Breeding and rangelands

The development of migrations, the systematic settling process of populations and the increasing mechanization have induced the development of a new type of breeding; that is semi-extensive and "sedentary". Indeed, the privatization of collective lands and the extension of cereal-producing areas following the use of tractors have significantly reduced the surfaces up to then reserved to grazing. Besides that, the fact that transhumance was no more effective in the region entailed the division of big herds into small flocks (from 20 to 50 heads).

Previously, collective shepherds who gathered animals belonging to various owners practiced a short transhumance, which enabled to use rangelands according to the season; these grazing lands were still collective and were located quite near the settling places (between 15 and 25 km). This type of herd management enabled vegetation to regenerate, in years of sufficient rainfall. The pressure was not continuous in the same places and the overall balance was not negative. But the sedentary flocks (2 to 20 heads) that resulted from the scattering of collective herds entailed the most severe consequences on environmental resources.

In fact, these herds, kept by old people, women and young children, are always grazing in the same places, in the neighborhood of houses, thus preventing any regeneration of the vegetation cover.

This results either in a strip of degradation and the disappearance of all vegetation (the crust is left bare in case of coherent formations), or in the very quick development of substitute vegetation species, of low pastoral usefulness (such as *Astragalus armatus* on crusts, and *Stipagrostis pungens* on sandy soils).

This type of landscape characterizes the bordering areas of the densest rural settlements, where sedentary herds are the most numerous, particularly in the *Medenine* plateau (Cf. unit 5 in Fig. 4).

In conclusion, the study of the environmental dynamics in the *Jeffara* region, between 1972 and 1985 evidences the consequences on the spot of the transformation of a *collective use* into an *individual use* of natural resources. In order to highlight the trends of the environmental dynamics, the solution implemented consisted in analyzing, synthesizing and simplifying original documents, by grouping together categories that offered little interest within the scope of this study, so that only the main themes were kept. We have then grouped the various categories of dynamics together, and reduced them to only five per date (1972 and 1985). As result, we have 25 possibilities of dynamics.

Of course, this grouping was performed after a visual exam of both classifications in 12 categories of 1972 and 1985, which showed that the process was more important in the sandy plain and in the area of olive groves. In conclusion, human factors remain the main ones, whereas physical and climatic constraints are only favorable factors in most cases.

From 1972 onwards, and given that some areas were already desertified and transformed into mobile dunes, new lands which were still intact and kept as rangelands were affected by a pressure in relation to the rate of allocation of collective lands.

This was the situation in the *Jeffara* region; we will now consider the *Nefzaoua* area.

5 Landscape dynamics in the *Nefzaoua* region

Just like in the *Jeffara* region, human factors are here too the main driving forces of the environmental dynamics. They are materialized by the unprecedented development of irrigated areas, the reduction of grazing lands and the Stalinization of soils. The *Nefzaoua* region differs therefore from the *Jeffara* area in its harsher climatic conditions and soils. In fact climate is characterized by a mean annual rainfall lower than 100 mm, with prevailing east winds that are often violent, high summer temperatures, and a mean ETP of 1,600 mm/year, which entails a quite permanent and acute water deficit.

However, the regional climate favors the cultivation of dates, mainly of the *Deglet-Nour* variety, providing good quality dates, at the condition that the trees are irrigated abundantly.

Arable lands are scarce; they are characterized by a constant artificial input of organic matters and minerals that may amount up to 40 cm deep. These soils are highly sensitive to salinization. Because of a bad drainage, they are soon saturated with water, which entails that salts concentrate at the surface.

In *Kebili* area for instance, the salinity of drainage water (assessed to be 981 l/s in winter) is comprised between 2.6 and 8.8 g/l. In summer, the amount of drainage water is less important (932 l/s) and includes more salt (2.9–12.3 g/l).

After this brief survey on physical conditions, let's analyze the main trends of the environmental dynamics in the *Nefzaoua*. Two periods may be distinguished, the first from the beginning of the century to 1985, and the second one from 1985 to date.

5.1 The 1905 to 1985 period

During that period, a first analysis of maps dating from the beginning of the century (1905) shows that the southern *Nefzaoua* region was very little populated, and that the landscape mainly included *Shotts* (playas) and *Dunes* (sand accumulations) (Fig. 6a).

This is why, excepting *Jemma, Douz and El Golea, Nouail, Zarsine* and *Blidet,* oases were only small islands of isolated palm-trees, created near sources, thus being used as stopping places and bivouacs on the route of caravans.

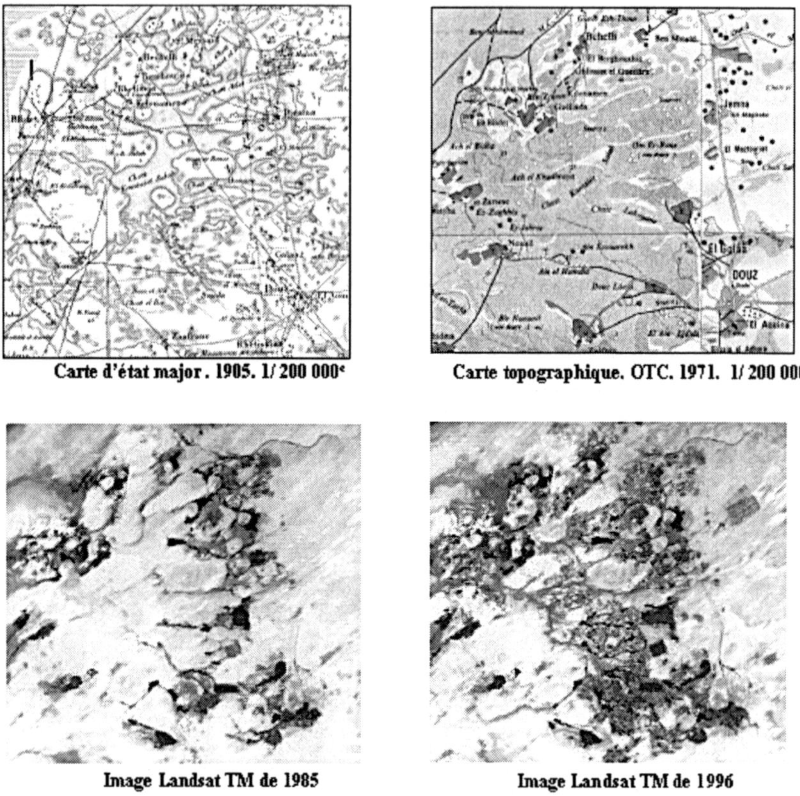

Fig. 6 (a–d): Localization of described landscapes

In fact, this situation rapidly evolved during the mid-1940s, with the increasing settlement of nomads and the introduction of new technologies for boring wells.

As direct impacts, the number of new oases dramatically increased around these places. However, it is worth pointing out that the creation of wells was always undertaken under the supervision and control of the state during this whole period.

As was the case for the Jeffara region, the 1960s were marked by the implementation and further abandonment of the "*collectivist*" period (1963 to 1969), as well as by the e*migration* to Europe of thousands of oasis workers, who used to look after irrigated areas.

In spite of the fact that new wells were still being bored all over the *Nefzaoua* region, no considerable land use changes occurred at the beginning of the 1970s.

The main causes were the quite generalized abandonment of plots during the "*collectivist*" period, as well as the departure abroad of a large part of the labor force from the oases. Aerial photographs of 1971 and the topographic map at 1/200,000 (Fig. 6b) and the *Landsat TM* satellite image of 1985) evidences a clear resumption of the setting up of new oases in the neighborhood of wells in the southern *Nefzaoua* area (Fig. 6c).

This resumption might be ascribed to the intensification of the Regional Development Programme (PDR), to the privatisation of collective lands, and above all to the new status of Kebili as a *Governorate* and lastly, to money transfers from regional workers employed abroad. However, the mid-1980s was the period that marked the greatest dynamics in the landscape of the southern *Nefzaoua* region, especially because of the rush to create *hand-drillings.*

5.2 The period between 1985 and 1996

Since the end of the 1980s and the beginning of the 1990s, the southern *Nefzaoua* area is subject to a considerable dynamics, mainly connected with the rapidly increasing number of individual wells and the creation of new irrigated areas around these works.

In fact, the *Landsat TM* image of 1996, either by direct reading (simulated colour image) or after processing (vegetation index, addition and removal of channels, classification,...) shows an unprecedented extension of new irrigated areas created between 1985 and 1996, especially from the *El Melah wadi* to *Douz*; most often west of the road from *Kebili* to *Douz* (Fig. 6d).

The biggest concentration of individual creations is located in the northern and western neighborhoods of *Douz* and *El Golea*, *West of Jemma*, around *Blidet*, *B'chelli* and *Zarsine*.

The analysis of the topographic situation of the new private creations shows that in most cases, or even in all cases, these new creations were undertaken on low, ill-drained plots, which are consequently easily saturated.

The situation is all the more worrying that private drillings are artesian wells, i.e. they operate non-stop and pour out water surplus (*Màa Zayid*) outside the irrigated area, in the lower zones nearby the irrigated area. Because the number of drillings rapidly increased (about 1,700 at present), the area of humid zones is extending as never before; they now represent "homogeneous" plots that cover large surfaces.

The analysis of the satellite image of 1996 shows that, near areas irrigated by drillings (which can be recognized by their specific spectral signature), there are a "gallery" vegetation (especially reeds) that grows along the water course to *Sebkahs* and that is colonizing more and more areas, as well as many "lakes" that results from enormous amounts of artesian water coming from hundreds of "hand-drillings".
In the southern Nefzaoua region, there is the following chain reaction due to this huge amount of water that pours out and then evaporates:

- Because there are so many artesian "hand-drillings" there is an unprecedented over-exploitation of resources and a quick drawdown of the water table.
- The drawdown of the water table induces an ever increasing pumping, that will be quite generalized around 2010 if the current situation remains unchanged.
- The surplus of irrigation water and the lack of drainage contribute to a rapid salinization of the soil in irrigated areas, and consequently to the degradation of their productive capacity, which is reducing every year till they become completely sterile, thus entailing predictable consequences on the life of populations.

- Another consequence that can be observed since the drawdown of the water table is the progressive withering of palm clumps, called « d'zira », that are dying in turn because of the lack of water. They are part of the landscape heritage, and have contributed to the originality and beauty of the Nefzaoua region from immemorial times. Now their decline will reduce the attraction to this area and will consequently affect the regional tourist activity.

In conclusion the recent extension of oases in the southern Nefzaoua region is usually uncontrolled and induces deep irreversible changes that affect water and soil resources as well as landscapes.

The current situation is characterized by a concentration of men and activities in a fragile environment. In order to avoid the risk of reaching the absolute limit of vital resources, it is essential to raise the issue of regulation and arbitration. It is also urgent to determine the dynamic bases for a viable environment, likely to generate a sustainable socio-economic development that should be sufficient to meet the increasing needs of the present population, without undermining or mortgaging the future.

Is sustainable development irremediably lost?

We will try to answer this question in the following chapter, by considering the driving forces of a sustainable development through the case study of the Sehib region. Indeed, the way resources are managed in this region has been based on local knowledge and has been successful for centuries. It appears to be one of the best, efficient and reasonable methods to manage resources and therefore is worth understanding, restoring and reinventing.

6 Restoring and reinventing sustainable natural resources management in the Sehib Area

In spite of the quick progress of modernization and economic changes, there still exist some traditional management systems and agricultural and pastoral knowledge that suit their environment. They rely on the use of local resources and are applied at a small scale, in a decentralized way; they favor the preservation of natural resources. Within this context, in spite of a high aridity and of human activities dating back to the Roman period, the Sehib region located in the low southern plains (Fig. 1) seems to be an exception for it kept a balance that contrasts with its immediate surrounding. In fact, such a system shows an unprecedented adaptation of men to their precarious environment. It is based on the knowledge that women and men of this region have been acquiring regarding their environment, and on the ways it determines their everyday life.

This local knowledge, whatever its form (crops, rotation techniques, practices, and technologies) is based on experience and is often inherited from previous generations; it has been tested century after century, and is still experimented to meet local conditions, needs and requirements.

Now, it is largely acknowledged that such farming strategies should be used as guidelines to design projects of research and development in rural environments.

In view of revealing aspects of ill known or little studied concerning arid regions and of drawing the attention of decision and development makers towards the need to restore and integrate this local knowledge into the process of research and development, we will rely on the study of the Sehib region.

Consequently, we will first highlight the various principles that govern such an adaptive strategy (which is more than 3,000 years old) in relation to environmental constraints. This policy mainly benefits from an ingenious know-how and a collective discipline adjusted to local conditions; resources are exploited in the best reasoned way and in view of their sustainability, as if they were a heritage rather than mere resources.

6.1 Tools and method

The method selected relies on scientific information as well as on local knowledge, as the main components of the process of data collection and information analysis.

In the following (Fig. 7), we will try to sum up local knowledge and scientific learning in order to understand how resources are managed in the Sehib region.

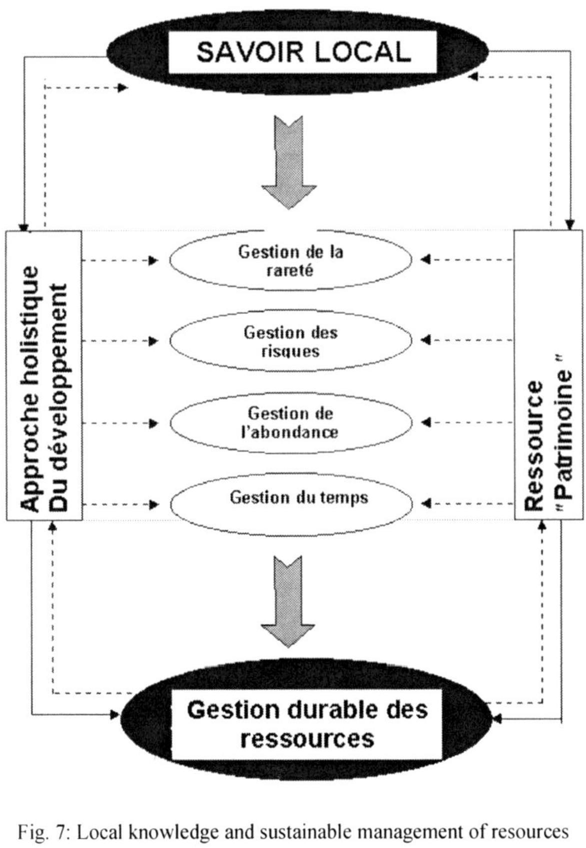

Fig. 7: Local knowledge and sustainable management of resources

This approach aims at integrating the sources of local knowledge as the main components of the process of data collection, information analysis and decision making.

Collecting information from the holders of local knowledge was undertaken with a representative sample of population, according to techniques of local diagnosis, derived from active methods of participative research and planning.

In the meantime, existing documents and aerial photographs of various dates were analyzed, and thematic information were processed and derived from recent satellite images. Lastly, a set of climatic data covering nearly a century was analyzed.

7 Basis and principles for an optimum management of resources

According to physical and climatic constraints, land uses vary from one year to the next, in relation to the rainfall recorded and above all, to the seasonal distribution of rains. The exploitation of lands mainly relies on the long-term experience of farmers, adapted to the physical conditions of the area. The system of land use entails no competition whatsoever between cereal crops and pastoral activities; it rather consists in a complementarily to meet the needs of farmers and herds.

Based on inter-annual variations of climatic conditions, three types of agricultural campaigns may be distinguished: favorable, average, and unfavorable campaigns.

If climatic conditions are favorable, i.e. if there is early abundant rainfall (in autumn or even at the beginning of winter), farmers go to plains in order to plough their lands and sow cereals, while herds are kept in the mountains during the winter season. The rains that occur in February and at the beginning of spring (above all in March) ensure cereals to give good production.

Unfavorable campaigns are characterized by very late rains, which, even though abundant (at the end of winter) are poorly efficient and do not allow to plough lands and to sow cereals. In that case, herds go to the plains where they spend the winter. At the beginning of summer, they come back to mountains and pastures. This enables to reduce the pressure on rangelands and to avoid a risk of overgrazing, which affects several regions of southern Tunisia.

7.1 Guidelines for the management of the "heritage" resource

Interviews with various generations of populations who hold local knowledge, field surveys, the compilation of existing documents and the exploitation of aerial and satellite documents enable us to draw a diagnosis of the main guidelines of local knowledge, as well as the means implemented to achieve an optimum resources management, within the scope of a holistic approach for a sustainable development (Fig. 7).

However, it is worth mentioning that this work is only the outline of a deeper research; which will allow to better take into account the elements and interactions of the system of local knowledge and to integrate them into the process of research and development.

In fact, the strategic line that governs local knowledge consists in considering a resource as a strategic heritage to be carefully managed, and not as a product to be exploited immoderately. This is why this management has to be implemented while considering current needs as well as those of future generations (Fig. 7).

This kind of highly emotional relation between population and their heritage requires an unfailing individual and collective discipline that highlights the interest of the social group, as well as all the interactions that link the various components of the environment, within the scope of a holistic approach.

This strategic concept of the "heritage" resource is implemented on the field through a set of guidelines that take into account climatic and soil conditions, as well as notions of time, scarcity and abundance (Fig. 7).

The local experience that has been acquired and passed down for centuries relies on the following principle: never try to force environment or to control nature, but adapt yourself. Hence, local experience has set up a series of guidelines that govern the relations between men and their "heritage" resource, namely including:

- Management of scarcity, that consists in knowing adequately the limits beyond which resources should not be used, in order to avoid risking their degradation.
- Management of time is tightly connected with the previous notion, since time is the factor that determines the degradation of the resource. This is how the time factor gave birth to the notion of mobility as a strategy to adapt to the scarcity of resources.
- Management of hazards concerns the adaptation to climatic uncertainties. On the field, it is materialized by diversifying activities (farming, breeding, trading), or dividing up the risks of a single activity (cereal crops) by doing it in different bioclimatic areas.
- Lastly, management of abundance means that the rare favorable or exceptional years are turned to good account. This namely consists in keeping part of a good crop in anticipation of difficult years (that are more frequent than favorable campaigns), in selling the bare minimum and only what is needed to diversify.

8 Conclusion

In the Jeffara and Nefzaoua regions, the various examples we have just described evidence that the environmental dynamics is showing a trend towards degradation. This dynamics of degradation is defined as "a process of aggression of the soil by different factors whose combination and permanence transform a territory that was initially fertile into a barren land, unsuitable for any production and likely to interfere with bordering areas by making them barren". Whatever the definition selected, degradation is usually induced by men, who disrupted the balance between them and their environment. Physical and climatic constraints are only favorable factors in most cases.

Technically, fighting against degradation seemed easy and feasible; but poor results all over the world showed that an *alternative approach* of the issue is vital, and that an *unfailing discipline* from local populations and long-term efforts from governmental authorities are necessary.

Within this scope, we have briefly outlined, through the example of the *Sehib* region, the relation between men and resources, and the importance of protecting and restoring local knowledge. This teaches us several lessons:

- Whereas *modern science* and development efforts aim at *controlling nature*, local knowledge on the contrary *is adaptive*. The societies that hold such knowledge gain an innermost knowledge and a good understanding of their own environment, that we have to grasp better.
- Nevertheless, the knowledge of rural populations is limited. It is often local, practical, empirical, but mostly intuitive. As such, it is quite a closed system that depends on what farmers can directly observe. Peasants have no access to theoretical knowledge or specialized techniques and therefore, cannot wholly access to outside resources.
- The failure of several projects for rural development raises the question of the inadequacy *of research* regarding the design and dissemination of technologies likely to be widely adopted by local populations, who own few resources and live in a less favorable environment.
- Consequently, the aim of research and development should be to foster the *symbiosis* between *local knowledge* and the innovations of the *official science.*
- The best way to reach this objective is to promote the whole *involvement* of farmers regarding the development of *suitable technologies.* This involvement is a prerequisite if we are to adapt exogenous technologies to the social, physical and ecological environment of the targeted region. But this involvement is also essential to ensure the necessary *feedback* of information to scientists. Farmers' involvement will be in fact a *corrective factor* for *mistaken assessments* done by developers or researchers who belong to the *official science* because their vision of agricultural production systems is not enough holistic (Fig. 8).

Lastly, it can be said that, now that the unavoidable transformation of the means of production entail changes all over the world, it seems that *traditional methods of farming and management are justified for the sustainability of both the human and ecological environments.*

Besides, the in-depth study of traditional agro-ecosystems allows designing new agro-ecosystems that have been improved to solve the various problems connected with the "modern" approach.

Local knowledge is the outcome of a cultural and biological evolution for many centuries; it is the *fruit of experience of interaction with environment accumulated by farmers who had no access to outside inputs, capitals or scientific knowledge.*

Relying on their *experience*, these farmers have set up sustainable agro-systems by using the *resources that were available* on the spot, as well as human and animal energy.

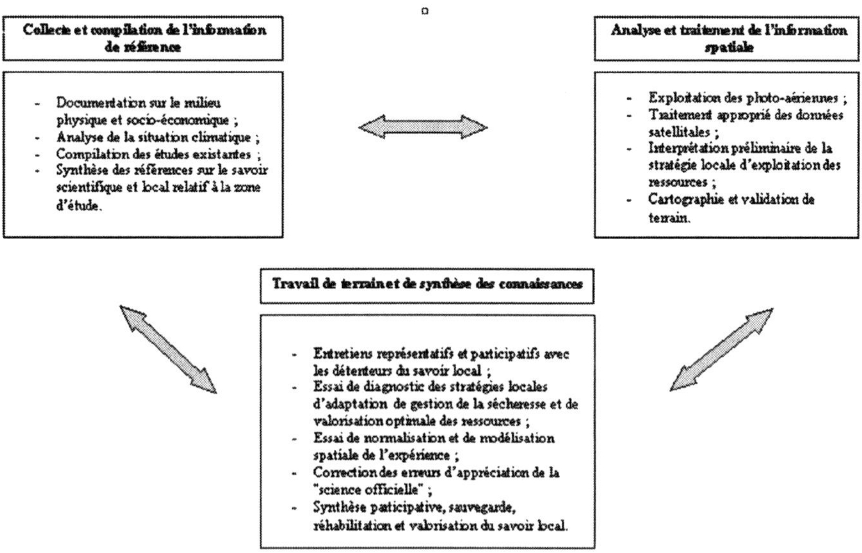

Fig. 8: Participatory approach for integrating local and scientific knowledge's for optimal resources management in arid mediterranean environment

Bibliography

Abaab A. 1981. *La marginalisation des techniques de la petite hydraulique familiale dans le Sud tunisien: cas des "Souanis" de Ben Gardane*, thèse de Doctorat de 3è cycle. Université Paris VII. 263 p.

Abaab A. 1984. *Aridité et transformations socio-économiques dans le sud de la Jeffara tunisienne*, Colloque de Géographie Maghrébine, Constantine (Algérie), 28–30 mai 1994. 37 p.

Benmansour N. 1999. *Le problème de l'ensablement dans la plaine de la Jeffara -Sud Est de la Tunisie. Etude par télédétection satellitale et observation de terrain*, Institut des Régions Arides. 32 p.

Bouhaouache T. 1983. *Les incidences des phénomènes migratoires sur les transformations des systèmes d'économie agricoles dans le Gouvernorat de Médenine*, thèse de 3ème cycle. Université de Montpellier 1. 169 p.

Bouhaouache T. 1986. *L'apurement des terres collectives dans le Centre et le Sud de la Tunisie*, in Revue des Régions Arides, numéro spécial (1991). Pp: 197–204.

Dregne H.E. 1977. *Status of desertification in the hot arid regions*. Conférence de Nairobi, 1977.

Floret C., Pontanier R. 1976. *Relation climat, sol, végétation dans quelques formations végétales spontanées du Sud tunisien. Production végétale et bilan hydrique des sols*. Document technique n 1, I.R.A., 1978. 96 p.

Floret C., Pontanier R. 1982. *L'aridité en Tunisie présaharienne: climat, sol, végétation et aménagement*. ORSTOM, Paris. 544 p.

J'Guirim H. 2000. *Tendances pluviométriques en Tunisie 1901-1990*, thèse de Doctorat en Géographie. Faculté des Sciences Humaines et Sociales. Université de Tunis I. 1200 p.

Khatelli H. 1981. *Recherches stationnelles sur la désertification dans la Jeffara (Tunisie): dynamique de l'érosion éolienne*, thèse de Doctorat de 3ème cycle. Université Paris I. 218 p.

Khatelli H. 1983. *Contribution à l'étude de la dynamique des dunes dans les parcours du Sud tunisien*. Séminaire IRA/UNESCO sur les problèmes de l'érosion éolienne dans les zones prédésertiques. Jerba, 21–26 novembre 1983. Pp: 21–33.

Khatelli H., Novikoff G. 1983. *Résultats préliminaires sur les travaux de lutte contre l'érosion éolienne dans les terres cultivées en céréales de la Jeffara tunisienne*. Séminaire IRA/UNESCO sur les problèmes de l'érosion éolienne dans les zones prédésertiques. Jerba (Tunisie) 21–26 novembre 1983. Pp: 103–106.

Louis A. 1979. *Nomades d'hier et d'aujourd'hui dans le Sud tunisien*. Editions EDISUD – 1979. 334 p.

Ministère de l'Agriculture 1972. *Séminaire de Gabès sur la lutte contre la désertification. 27–29 décembre 1972*. Rapport de synthèse, Gabès. 24 p.

Mzabi H. 1988. *La Tunisie du Sud-Est: géographie d'une région fragile, marginale et dépendante*, thèse de Doctorat d'Etat. Université de Tunis I. 2 tomes. 941 p.

Nasr N. 1993. *Systèmes agraires et organisations spatiales en milieu aride: cas d'El-Ferch et du Dahar de Chenini – Guermessa*, thèse de Doctorat en Géographie. Université Paul Valery, Montpellier III, 1993. 217 p.

Novikoff G. 1983. *Essai de lutte contre l'érosion éolienne dans les parcours à Rhanterium suaveolens de la Jeffara et leur application*. Actes du séminaire IRA/UNESCO sur les problèmes de l'érosion éolienne dans les zones prédésertiques. Jerba (Tunisie) 21–26 novembre 1983. Pp: 97–102.

Omrani S. 1982. *Le territoire des Benizid (sud tunisien): modes de production et organisation de l'espace en zone aride*, thèse de 3è cycle. Université Paul Valery. Montpellier III. 375 p.

Rapp A., Le Houerou H.N., Lundholm B. 1976. *Can desert encroachment be stopped? A study, with emphasis in Africa*. Report published in cooperation between the United Nations Environnemental Program (U.N.E.P) and the Secretariat for International Ecology. Stockholm. n.**24**. 241.

Talbi M. 1983. *Dégradation d'une steppe à Rhanterium suaveolens dans le sud tunisien* en collaboration avec A. El Hamrouni. Actes du séminaire IRA/UNESCO sur les problèmes d'érosion éolienne dans les zones pré désertiques. 3 p.

Talbi M. 1993. *Contribution à l'étude de la désertification par télédétection dans la Jeffara (Sud-Est tunisien)*, thèse de Doctorat en Géographie. Université de Tunis I, 1993. 2 tomes. 305 p. 54 cartes et Figs. 8 planches en couleurs.

Talbi K. 2000. *Intégration des données multisources pour l'étude de la dynamique des paysages arides: CAS des oasis de Fatnassa/Béchri (presqu'île de Kébili)*. Mémoire de fin d'études d'Ingénieur. Laboratoire de Télédétection et de l'Information Géographique. IRA. Medenine/ESIER. 75 p.

Talbi M., Nasr N., Benmansour N., Talbi K. 2000a. *De la gestion collective à l'exploitation individuelle des ressources en eau et ses conséquences sur le milieu saharien: Cas des oasis du Nefzaoua méridional. Tunisie*. Actes de Séminaire International MEDENPOP 2000. Octobre 2000. Jerba.Tunisie. 4 p.

Talbi M., Guesmi N.J., Guirim H., Benmansour N., Talbi K. 2000b. *Savoir local et gestion optimale des ressources en environnement aride: Cas de la région du SEHIB. Tunisie*. Actes de Séminaire International. MEDENPOP 2000. 8 p.

Talbi M., Benmansour N., Talbi K. 2000c. *Vers une utilisation optimale des moyens modernes d'investigation pour le contrôle et la gestion du phénomène de la sécheresse et de la désertification dans le Monde Arabe*. (en langue Arabe), in Journal Syrien de Télédétection. n.**12**. November 2000. Pp: 24–37.

QUALITY MANAGEMENT IN INFORMATION SYSTEMS FOR COMBATING DESERTIFICATION AND DRY LANDS MANAGEMENT

MOUNIR SELLAMI[1] AND MOHAMED TALBI[2]
[1]*Sahara & Sahel Observatory (OSS), Tunisia, mounir_sellami@yahoo.fr*
[2]*Institut des Régions Arides, Medenine, Tunisia, Mohamed.talbi@yahoo.fr*

Abstract

As shown in the Convention to Combat Desertification (UNCCD), as well as in Chapter 12 of Agenda 21, desertification is one of the world's most serious environmental problems, threatening the earth's fertility and population food security.

Desertification is spreading due to climatic causes and human activities, both those performed in a single area and those performed all over the world (climate change). Some measures can only be taken at a general level, but others have to be taken at regional, national or local level. Combating desertification and the tasks to mitigate the effect of droughts requires an integrated treatment of the physical, biological and socio-economic aspects.

Understanding natural, social, cultural, economic and political features of local and national environments, in particular their dynamics and interactions, is considered as one of the basic principles of the Convention to Combat Desertification.

Improvement of the knowledge base is widely considered as a necessary first step for addressing the relevant problems and developing appropriate solutions. Many research and data-collection activities are being implemented worldwide, mainly to better understand the state of the natural resources in dry lands areas.

Such dry lands cover 40% of the world's land surface, and are the habitat and source of livelihood for more than one billion people. Of the 5,200 million hectares of dry lands used for agriculture, more than 70% are suffering from degradation. Africa, Asia, and Latin America are particularly under threat, but some 30 million hectares of European territory, especially those bordering the Mediterranean, are also potentially affected, threatening the livelihoods of over than 16.5 million people.

The mind of partnership as inspired by the UNCCD requires flexibility and simplicity in information circulation to all levels of decision: from local to international, and vice versa.

The assessment of the current status of desertification in a country is a crucial first step for understanding and correcting the problem. National assessment maps are needed for planners and decision-makers to establish priorities for combating desertification. Considered the cost of the environmental damage caused by land degradation and considered the absence of reliable information, sustainable decisions can hardly be made.

Dry land desertification can be remedied or even reversed, if updated information is provided on affected areas; especially concerning land use detection.

A. Marini and M. Talbi (eds.), *Desertification and Risk Analysis Using High and Medium Resolution Satellite Data*,

Satellite images can highlight relevant land use change along with increased surface reflectivity, temperature, dryness and dustiness; infrared sensors can detect vegetation stress due to environmental shifts.

This satellite data, if combined with in-situ information, processing tools, models, databases management and geo-information systems (GIS), can create standardized and comparable geo-information products, which can also be used to satisfy UNCCD reporting requirements.

Advanced technologies for helping desertification assessment and monitoring, can be of great value: computers, satellite images, GIS and databases management systems have revolutionized natural resource data collection methods. However, these tools do not substitute field studies.

Another problem related to gathering data on arid lands is the diversity of sources, which influences negatively, and sometimes crucially, the GIS and databases quality. As result, the quality of products used in making decisions, can be more or less useful or erroneous. Therefore a quality control of the natural resources management databases in the arid regions becomes a necessity.

Keywords: Environmental databases, desertification information systems, data quality, arid lands management

1 Introduction

The quality management is an essential component of all types of projects. Indeed, it permits to ensure that the project is going to be able to satisfy the needs for which it has been undertaken. For a project concerning a geographical information system, we will show that the importance of the quality management is exacerbated because of the specificities of a G.I.S. In this setting, we will try to insist especially on these specificities, and the generalities will be reduced to the minimum. In a first time, we will recall a few basic principles and definitions. Then, the focus will be successively in the three components of the quality management: quality planning, quality insurance, and quality control, while the chapter will show every time how each component applies in a G.I.S project.

2 The quality principle

In order to better formalize the concepts and to have a common universally recognized referential, some norms have been established to give precise definitions of the different terms used: these are the ISO norms. Thus, the norm ISO 8402:1994 indicates that: The quality is "the set of the features of an entity that confers him the faculty to satisfy needs either expressed or implicit."

Rephrasing it, we can consider that the quality of a GIS project is what allows it to produce valuable results, knowing that the value is a "judgment carried on the product on the basis of the needs and the user's incentives, expressed by a size that increases when the satisfaction of the user's need and/or the expense of the product decreases" (NF norm X 50–150). From that definition, we can easily deduct that quality management in a project interests both the methodology and the product.

According to the norm ISO 8402:1994: "The quality management includes all general activities of management that determine the quality politics and objectives, the responsibilities, and sets them in motion by the quality planning, the quality insurance, and the quality control" (Fig. 1).

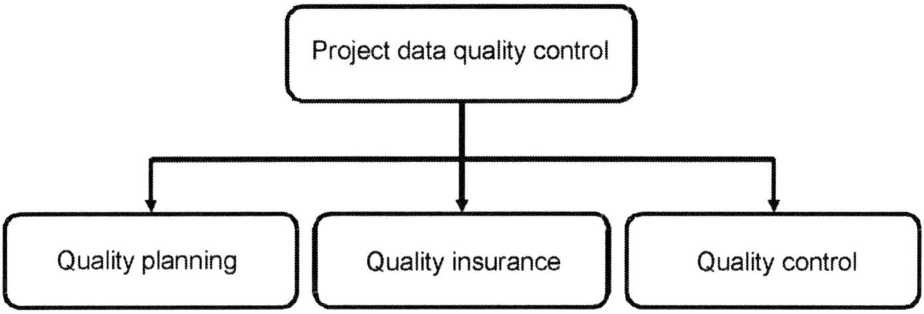

Fig. 1: Activities of quality management

2.1 Quality planning

Quality planning tries to identify the quality criteria that are important for the project and the way in which it will be possible to satisfy them.

Whatever the project is, these criteria must be defined since the beginning, and must concern the used procedures, as well as the final products of the project. They depend on two main elements: the quality politics within the organism, and the description of the product. From their survey, we will be able to get different useful documents for the management of the project quality.

2.1.1 The policy of quality in the organism

The policy of quality is "the set of orientations and general objectives of an organism with regard to the quality, as they are expressed positively by the general direction."

If it exists in the organism, this policy of quality can be adopted directly and used in the project. Otherwise, the chief of the project will be in charge of defining them. In any case, it is particularly important for all the actors in the project to be perfectly conscious and aware of this policy.

Thus, project collaborators must clearly have understood why some criteria of quality must be applied. Indeed, if the policy of quality is imposed by force without any explanation, it is likely that there will be some resistances and that the rules will be shaped.

2.1.2 The descriptive project

During the phase of needs for evaluation, the project extent has been defined. In particular, the features of the final product have been specified, and it is obvious that these technical specifications have some repercussions on quality planning. In our case, the final product is the supply of benefits around a G.I.S. This implies the existence of many criteria to be respected during the project.

For example, the choice of the precision of the data that enters in the G.I.S. can be an important task. Conventions to name the represented geographical features should also be adopted; data classification should be described in detail and should be applied, etc. Evidently, the criteria will not be arbitrary, but will be adapted to every project. If imposed without particular reasons of precision, high imperatives or elements exhaustiveness of the database entails a strong increase of the time and money required.

On the contrary, a too weak precision in the data risks preventing the G.I.S. to answer to the questions for which the project had been formulated. It is therefore imperative to have always in mind the project ultimate objectives, which permits to define the criteria of quality in a better manner.

2.1.3 Results of quality planning

The set of the previous considerations must be able to permit a definition of a quality insurance plan. This plan describes how the project management team is going to set in motion its policy of quality. According to the ISO 9000 norm, it consists to describe the "quality system" of the project.

According to ISO 8402:1994, the quality system is "the whole organization, the responsibilities, the procedures, the processes and the necessary tools to set in motion the quality management".

This quality management plan must firstly indicate notably how the different stages that will be reviewed subsequently will take place, in order to know quality insurance, its control and its possible improvement.

Secondly, this phase of quality planning must permit to get operational definitions. It consists to describe, with precision, activities terms at every stage, procedure or process covers, and how the efficiency of each of these activities will be measured at the time of the quality control. In the same way, control lists (checklists) can be realized in order to verify that the stages of a process have been well achieved.

3 Quality insurance

Quality insurance corresponds to "the set of the activities pre-established and systematic setting in work in the setting of the quality system and demonstrated as need, to give the confidence appropriated in what a product will satisfy to the requirements relative to the quality" (ISO 8402:1994).

It consists, therefore, to formalize the tools put in place in order that the quality of the products and services does not leave anything to the case and it is maintained and improved at time.

The quality insurance answers an obligation of means. It takes place along the project duration and at the same time aims at internal and external objectives: within the organism, it gives to the direction confidence in the project; in the commercial situations, it reassures customers.

Furthermore, we expect that more and more customers will require to their suppliers guarantees of quality insurance. Currently, quality insurance often passes by the respect of some models in relation to which the organism can be certified. In the following, we will start with brushing the big lines of these models, and how they can be applied in a G.I.S project.

3.1 The modes of quality insurance

The family's norms ISO 9000 are the result of an international consensus on the good practices of the quality management.

It describes a set of generic and normalized requirements to get a system of management with quality efficiency. These norms can be applied to all organizations and whatever is the product or service.

Indeed, ISO 9000 refers to the manner of working of the organization and not directly to the work result. In other words, the quality insurance is interested in the quality procedures, and not in the products ones. It consists, above all, in giving confidence in the fact that the organization masters its methods of work.

The family of norms ISO 9000 contains three models to which the organism can be certified: ISO 9001, ISO 9002 and ISO 9003. These three models distinguish themselves by their domain of application:

ISO 9001 establishes requirements relative to activities from the conception and the development to the installation and to the associated benefits.

ISO 9002 does not take care of the conception and the development, the other requirements being identical.

ISO 9003 concerns the needs of an organization that have resort to the controls and the tests to assure that the products and the final services satisfy to the specified requirements.

In this guide's setting, it will not be possible to review the requirements imposed by these models. In a general manner, we can say however that these models especially insist on the documentation of the procedures and products by the constitution of quality reports. This documentation can be made according to the plan of quality management and the operational definitions that could be defined in the phase of quality planning. The final version of these norms insists less on the importance of the "paper" putting the customer in the centre of the quality system.

The concordance between the real techniques of work and the requirements of the norms can be valued through audits quality, other tools of the quality insurance. These audits can be achieved by internal listeners or by independent organisms, what allows the organism audited to take advantage of a certification ISO 900x. In France, the obtaining of a certification passes by a demand to the AFAQ (French Association for the Quality Insurance).

In Tunisia, the Topography and Cartography Office (OTC) and National Remote Sensing Centre (CNT) are expected to be in charge of assigning the qualify level of the norms respect.

3.2 Application in a G.I.S project

All procedures previously described can be applied to the project management whatever it is, in particular in a G.I.S project.

Nevertheless, it is worth to first notice that by definition, a project is temporary: it risks of not being long enough to harvest the results of the quality improvement. The quality insurance is, therefore, to be seen as a necessity on the long term that concerns the set of the organism activity. Once implemented, the system will have positive repercussions on the set of the projects led in the organism.

Secondly, it also seems crucial to insist on the documentation in a G.I.S project. Indeed, a G.I.S project is not to be under-evaluated: the implementation system is probably to be in place for several years. It should be able to evolve with the time, the initial creators being not necessarily present. In a convenient view, all procedures and the operative modes must be retailed and consigned. For a G.I.S project, we can retail how the geographical data have been gathered, from which sources and how they were geo-coded. Therefore the documentation is essential: it contributes to the project value, and will allow the future use of the product.

Finally, we can give an example of quality insurance procedures while considering the stage of the data entering, the same in many G.I.S projects.

To reduce the semantic mistakes, it can be interesting to imagine, since the database conception, some automatic mechanisms of control and correction. Thus, it is often possible to define precisely the type of every field, notably if it must be numeric and have the date format. It can be also useful to arrange ergonomically the fields to fill in, in order to avoid bad manipulations to the operator's mouse. Using these simple procedures, the operator will have less risk to mistake two fields. In the same way, the topology mistakes can often be corrected automatically at the time of the data entering by some programmed "reflexes" or predefined in the software.

4 The quality control

In relation to the quality insurance that was an obligation of means, the quality control is an obligation of results:

"The control quality consists in comparing the results of the measures to the requirements specified in order to determine if the conformity is gotten by each of the features" (ISO 8402:1994) (Table 1).

Table 1: Appreciation table of the quality of the information system

Function	Criteria of appreciation	Appreciation scale
Providing the necessary information to assist in decision making	Adequacy of the content of the Conceptual Model of Data (CMD), to the nominal land	Percentage (%) of the geographical concepts described in the CMD
	Adequacy of the quality of the data to the criteria of Quality Insurance	Assessment of the loss of information quality following the different procedures of treatment and/or of the non validated data
	Adequacy of the information system structure to the expressed needs, after the last revision	Rate of satisfaction of the needs in the operational and decision making levels

The controls concern the different intermediate processes, the final product, as well as the project management. For example, the evaluation of the delays and costs enters in the setting of the quality control. At the time of the development of a G.I.S., different elements of quality control must be considered. Of course, these strongly depend on the type of G.I.S project that has been developed and the considerations given below will not be all met according to the cases. Three elements of a G.I.S are susceptible to be controlled: the software, the database and the staff.

At the time of the pilot project survey, or at the time of the final setting up of the G.I.S., it is essential to ensure that the central software of the G.I.S. has reached the required criteria of quality.

The software can be a commercial version to which an especially programmed module, or specific software have been added. However, in any case, quality control expresses itself in terms of software ergonomics, performance, evolution, and the presence of the required functionalities .... Different tests in real situation should be done to verify the efficiency of the system on each of these points. In particular, it is necessary to keep always in the mind that the G.I.S. forms a complex system, and that it is not obvious that the different modules, material and software are well compatible.

4.1 Quality controls of the database

An essential stage of a G.I.S project is the constitution of the used database. This database can be created using of the documents provided by the customer that will be digitalized, or by the purchase of numeric maps from specialized organisms as the O.T.C. In any case, it is recommended to do the quality controls on the data and their incorporation in the database, because different problems can occur (Fig. 2).

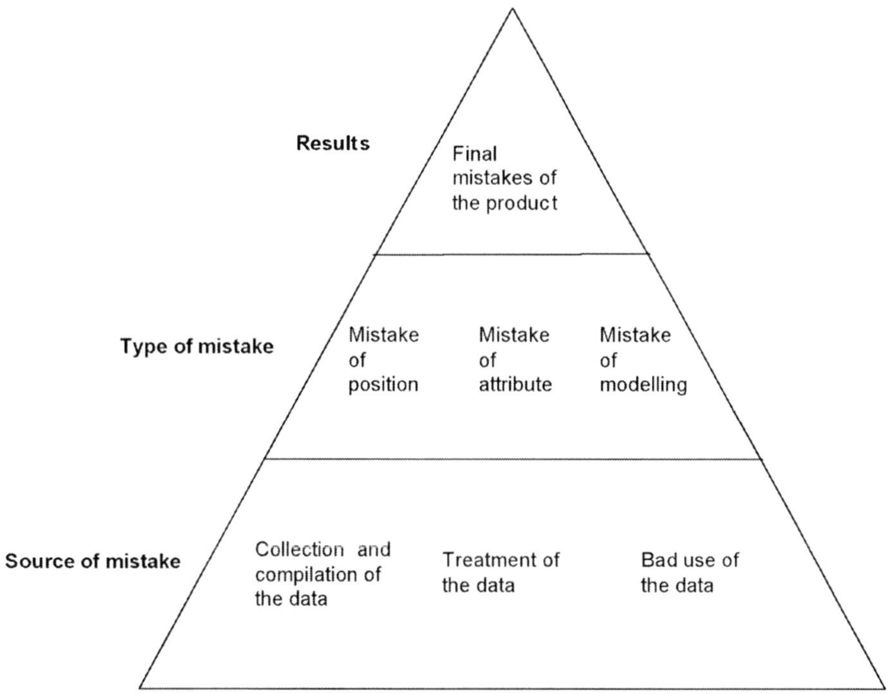

Fig. 2: Data monitoring strategy

4.1.1 Non-exhaustiveness

It is about verifying the conformity of the presence or the absence of the database elements in relation to the nominal land (picture of the universe through the filter of the definite specifications). Verifying concerns the presence of the specified data, the absence of that doubloons, etc.

4.1.2 Geometric mistakes

Two types of such mistakes are susceptible to appear: the relative positioning mistakes and absolute positioning ones.

- The relative precision is a measure of the maximal deviation of the distance between two objects on the digital map compared to the real distance on the land between the two same objects.
- The absolute mistake is a measure of the maximal deviation between the localization of an element on the digital map and its real localization on the land.

In order to correct this type of mistakes, the quality control must consist in regularly comparing the data to the reality. It implies to verify on the land the positions of the different elements, or at least to compare the data with sources of higher quality.

4.1.3 Semantic mistakes

In a G.I.S database, the attributes on the geometric data can be also erroneous. Then it is the accurateness of an object characteristic and not its localization that is considered: was a fast way confounded with a freeway, the operator having not selected the bad element in the software menu? As underlined previously, the automatic devices of the software and good ergonomics can permit, since the entering, the reduction of the occurrences of these mistakes, even though the quality control remains indispensable.

4.1.4 Topological mistakes

It consists to maintain the database integrity, which means to keep its topological consistency: we must try to detect the problems of polygons closing, non-connection between bows that should be linked... Most of the time, G.I.S software on sale contains tools that permit to do this control, sometimes in an automatic manner and in real time.

4.1.5 Data actuality deficient

It is necessary to always keep in mind that the geographical data are not unchangeable. They evolve in time, notably under the action of the natural elements (erosion, storms, etc.) and of the human activities (new constructions ...). The control of the data actuality is a necessity, especially as the efficient use of the G.I.S. is sometimes later than the data entering.

4.2 Training, qualification and authorization of the staff

A G.I.S project can drag a deep change of the methods and habits of work within the structure of the organism. Often, the staff that is going to use the G.I.S. has false or vague ideas on its possibilities. To maintain a good quality of the G.I.S results, some specific training will be necessary in various domains as statistical controls, sensitivity to the mistakes, precision evaluation ...

The staff in charge of the G.I.S should be sensitized to respect quality criteria, so that this becomes a natural element of their work.

For example, some updates must be regularly made or some elements must be corrected. The operators who will enter the new data should make it in the setting of a quality control process implemented since the beginning of the G.I.S use.

5 Conclusion

Monitoring and assessing desertification indicators are required if the effectiveness of control activities is to be measured and changes made, where necessary. Monitoring and assessing, unfortunately, have to be a long-term activity if they are to provide needed information on land degradation, trends that will not be easy to obtain.

In this way, more expressive and better informative products will be available for a more targeted decision-making, in order to better manage the problem of natural resources degradation in the arid regions, if the data quality control is conceived as a part of the information system.

A quality assurance (QA) program is a crucial aspect of the GIS implementation process. To be successful in developing reliable QA methods, individual tasks must be worked out and documented in detail. Data acceptance criteria constitute a very important aspect of the information system, and can be a complex issue. A full analysis of accuracy and data content needs will facilitate the creation of documentation, which may be utilized by the accuracy assessment team (internal) and constitutes a transparency proof.

A combination of automatic and manual data verification procedures is normally found in a complete QA program. The actual process normally involves validation of the data against the source material, evaluation of the data's utility within the database design, and an assessment of the data with regard to the standards established by the organization handling the information system conception, development and implementation.

A scheme for tracking packets of documents, as well as the resulting digital files is needed. This scheme should be able to track the digital files through quality control processes.

In such way, and with the improvement of the products quality (statistical and cartographic databases) and the whole information system, decision-making can reach a higher level of spatial and temporal precision (Fig. 3).

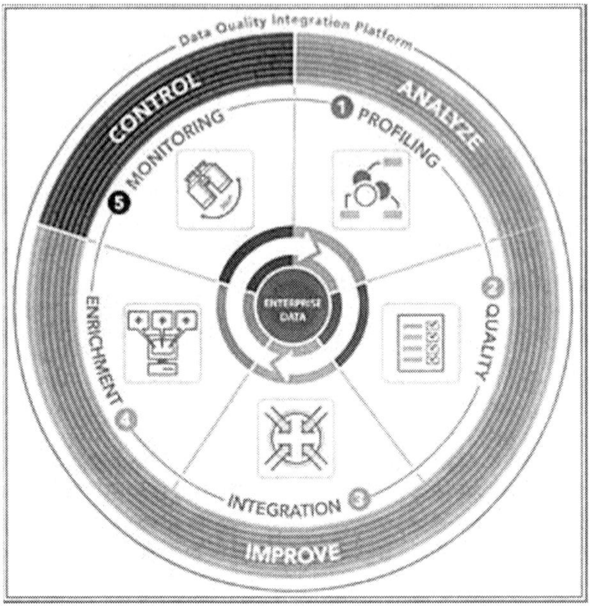

Fig. 3: Data quality integration platform

5.1 Definition of an investigation grid of a GIS Unit activities

The following lists present the different thematism and activities to study with the purpose of better knowing the state of the art as well as the perspectives of the GIS activities.

5.1.1 Analysis of the existing state of the art of the GIS activities

General objectives
Historic of the GIS
Main Constraints
Difficulties met
Equipment: hardware/software
Number of work stations and type of machines
Software of: GIS, remote detection, cartography and drawing. Name of software and number of licenses
Numeric data
Themes
Sources
Descriptions
Formats

Maintenance and update
Condition of use
Organization of the GIS activities
Services concerned
Staff involved
Functioning of GIS activities
Budget

5.1.2 GIS strategy

Perspectives of development
Means term
Short term
Existing cooperation
With organisms at the national level
With other organisms
Cooperation considered
With organisms at the national level
With other organisms

Bibliography

Anderson, J.R., Hardy, E.E., Roach, J.T., and Witmer, R.E., 1976, *A Land Use and Land Cover Classification System for Use with Remote Sensor Data*. U.S. Geological Survey, Professional Paper, Reston, VA.

Edwards, G., and Lowell, K.E., 1996, *Modeling uncertainty in photointerpreted boundaries*. Photogrammetric Engineering and Remote Sensing, 62(4), 377–391.

Ehlschlaeger, C.R., and Goodchild, M.F., 1994a, *Dealing with Uncertainty in Categorical Coverage Maps: Defining, Visualizing, and Managing Errors*. Proceedings, Workshop on Geographic Information Systems at the Conference on Information and Knowledge Management, Gaithersburg, MD, 86–91.

Ehlschlaeger, C.R., and Goodchild, M.F., 1994b, *Uncertainty in Spatial Data: Defining, Visualizing, and Managing Data Errors*. Proceedings of GIS/LIS, 246–253.

Pang, A., and Furman, J., 1994, *Data quality issues in visualization*. SPIE Proceedings on *Visual* Data Exploration and Analysis, 2178, 12–23.

Shi, W., 1997, *Statistical modeling uncertainties of three-dimensional GIS features*. Cartography and Geographic Information Systems, 24(1), 21–26.

REMOTE SENSING OF AEOLIAN DUST PRODUCTION AND DISTRIBUTION

KEVIN WHITE
Department of Geography, The University of Reading, UK
k.h.white@reading.ac.uk

Abstract

Airborne dust is of concern due to hazards in the localities affected by erosion, transport and deposition, but it is also of global concern due to uncertainties over its role in radiative forcing of climate. In order to model the environmental impact of dust, we need a better knowledge of sources and transport processes. Satellite remote sensing has been instrumental in providing this knowledge, through long time series of observations of atmospheric dust transport. Three remote sensing methodologies have been used, and are reviewed briefly in this paper. Firstly the use of observations from the Total Ozone Mapping Spectrometer (TOMS), secondly the use of the Infrared Difference Dust Index (IDDI) from Meterosat infrared data, thirdly the use of MODIS images from the rapid response system. These data have highlighted the major global sources of dust, mist of which are associated with endoreic drainage basins in deserts, which held lakes during Quaternary humid climate phases, and identified the Bodélé Depression in Tchad as the dustiest place on Earth.

Keywords: Dust, remote sensing, TOMS, Meteosat IDDI, MODIS

1 Introduction

Dust can be defined as mineral particles with diameters less than 500 μm, transported in suspension by turbulent air movements (Livingstone and Warren, 1996). The erosion, transport and deposition of this material creates hazards when it affects human activities. These are manifested at different scales. Locally, dust storms result in low visibility, creating hazards for road and air transport (Buritt and Hyers, 1981). There are also health concerns about the effects of airborne dust (Leathers, 1981; Wheaton, 1992) and the severe environmental impact of dust erosion became evident in the 'dust-bowl' years of the late 1039s in SW USA (Worster, 1979). Globally, there are significant uncertainties associated with the role of dust in the radiative forcing of climate (Fig. 1). Even the direction of forcing (whether cooling or warming) is unknown. Dust also has an impact on the nutrient budget of oceanic ecosystems, by providing a source of Fe to fertilize ocean regions that would otherwise be nutrient-starved (Liss and Turner, 2000).

A. Marini and M. Talbi (eds.), *Desertification and Risk Analysis Using High and Medium Resolution Satellite Data*,

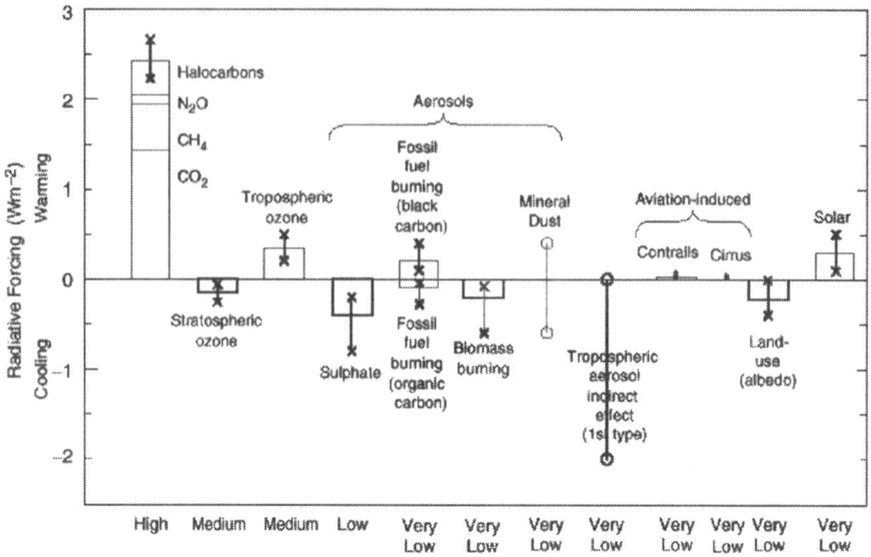

Fig. 1: The level of scientific understanding of radiative forcing processes according to the Intergovernmental Panel on Climate Change. In the case of mineral dust, even the direction (positive or negative) of the forcing is unknown (IPCC 2001)

The export of dust from desert regions such as the Sahara is well documented (Goudie and Middleton, 2001), and efforts have been made to quantify it (Ganor and Mamane, 1982). Deflation of superficial materials results in winnowing of fine material for export as dust, leaving a residual armoured 'stone pavement' surface typical of desert and peri-desert regions in drylands, though pedogenic accretion of aeolian material is also thought to be important in stone pavement formation (McFadden et al., 1987).

Vegetation is known to protect the surface from deflation (Wasson and Nanninga, 1986) and removal of vegetative cover is a documented cause of increased atmospheric dust loading (Leys, 1999). However, the relative importance of human activity versus climate variability in increasing dust flux to the atmosphere is debated (Middleton, 1985; Brooks and Legrand, 2000). There is some evidence to indicate a link between rainfall variability and dust storm activity, but the relationship is not always clear (Middleton, 1986; Middleton et al., 1986; Goudie and Middleton, 1992). Atmospheric turbulence is required to lift dust high enough into the atmospheric column to allow large-scale transport (Washington and Todd, 2005). Major atmospheric dust events are usually associated with strong seasonal winds (such as the 'Ghibli' seasonal winds in the Sahara), but dry convection can pick up a lot of dust (Cakmur et al., 2004). Surface topography can play a key role in the generation of the erosive winds by funneling; the Tibesti and Ennedi mountains create between them the Bodélé Low Level Jet (Washington et al., 2006b).

A key goal in dust research is the development of numerical models capable of simulating dust emission, transport, and deposition, so that the influence of dust on the climate system can be quantified (Washington et al., 2006a). However, much more detailed understanding of dust sources and processes of entrainment are required before this goal can be achieved (McTainsh, 1999). Remote sensing is proving to be a vital tool, by providing detailed observations of the desert dust cycle (Chiapello and Moulin, 2002).

2 Remote sensing of atmospheric dust

Satellite remote sensing can provide long time series of observations of atmospheric dust, which can help evaluating the importance of different sources (Washington et al., 2003). Attempts to quantify the global dust budget and characterize the main sources have involved three main remote sensing methodologies. Firstly, using data from the Total Ozone Mapping Spectrometer (TOMS) (Herman et al., 1997), spectral indices sensitive to atmospheric dust can be calculated (Torres et al., 2002). Secondly using METEOSAT infrared data, an Infrared Difference Dust Index (IDDI) (Legrand et al., 2001; Leon and Legrand, 2003) can be calculated. Thirdly, using data from the Moderate Resolution Imaging Spectroradiometer (MODIS) specific dust sources can be identified (Kaufman et al., 2005).

TOMS has proved to be one of the most important instruments for monitoring atmospheric dust. It is currently carried as part of the payload of a satellite mission called Earth Probe. TOMS is a Visible/UV spectrometer designed to make measurements of UV-absorbing aerosols in the troposphere. It can distinguish between aerosol types such as smoke, dust and sulphates. The capability to measure mineral dust from the TOMS instrument was originally developed using data from previous TOMS instruments which were carried on board in the Nimbus 7, Meteor 3 and ADEOS satellite missions. The Aerosol Index is based on an aerosol optical thickness retrieval algorithm, using observations at 340 and 380 nm by the Nimbus7-TOMS sensor, and to 331 and 360 nm measurements by the Earth Probe instrument. For the sake of continuity, however, the optical depth record is reported at 380 nm for both instruments (Torres et al., 2002).

EarthProbe TOMS was launched on July 2nd 1996, into a 500 km orbit, lower than the originally planned 950 km orbit, in order to decrease the size of the "footprint" of each measurement, and thereby increase the resolution and also the ability to make measurements over cloudless scenes. The lower orbit precluded full global coverage in the equatorial region (a role taken by ADEOS TOMS mission), but was designed to enable full coverage over the poles for monitoring the Antarctic Ozone Hole, which was a major objective of the Earth Probe TOMS mission. However, when ADEOS TOMS failed in June 1997, the orbit of Earth Probe TOMS was boosted to 740 km and circularized to optimize global coverage. Thus the mission, in its current configuration, is ideally characterized for monitoring dust transport, the majority of which takes place at lower latitudes (Middleton et al., 1986).

The TOMS Aerosol Index record (Fig. 2) covers the periods from January 1979 to April 1993 (Nimbus 7-TOMS observations), and from July 1996 to December 2000 (Earth Probe TOMS measurements). Files of monthly average optical depth on a 1 degree by 1 degree resolution grid can be downloaded from http://toms.gsfc.nasa.gov

Analysis of the long time series of TOMS observations has highlighted the major global dust sources (Washington et al., 2003). These are the Sahara, Middle East, Taklamakan, southwest Asia, central Australia, the Etosha and Makgadikgadi basins of Southern Africa, the Salar de Uyuni (Bolivia), and the Great Basin (United States).

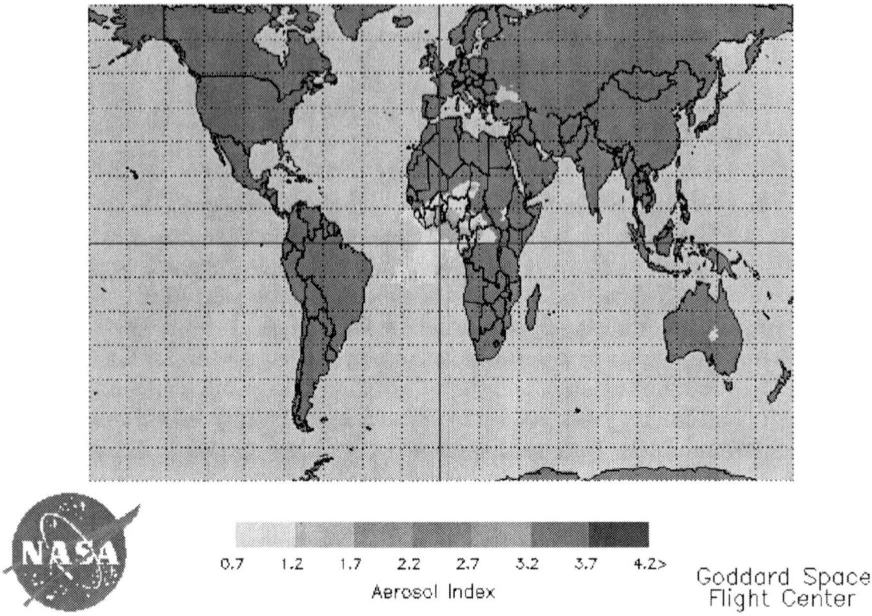

Fig. 2: Example Earth Probe TOMS monthly average Aerosol Index image for January 2000. See text for discussion (http://toms.gsfc.nasa.gov)

Analysis of Digital Elevation Model data demonstrates that most of these regions are large basins of internal (endoreic) drainage (Washington et al., 2003). Within the Sahara, two major source areas are identified (Fig. 3), firstly the Bodélé depression and, secondly, an area covering eastern Mauritania, western Mali and southern Algeria (Middleton and Goudie, 2001).

The Infrared Difference Dust Index dataset was developed at the Laboratoire d'Optique Amosphérique at the Université des Sciences et Technologies de Lille, France (Legrand et al., 1994). The Index characterizes the reduction in the measured infrared brightness temperature of the atmosphere from that which would result from a dust-free atmosphere. Brightness temperature values are retrieved from Meteosat infrared channel data collected at approximately midday. The maximum brightness temperature over a 15 days period is used to represent dust-free atmospheric conditions, and the daily brightness temperatures are subtracted from this, following masking out of cloudy pixels. The resulting index values are re-sampled to a 1 degree by 1 degree grid (Fig. 4). The IDDI signal results from the reduction in the temperature of the underlying land surface by reduced solar insolation, and also the attenuation of the outgoing long wave radiation by the aerosol layer. As a result, the IDDI method can only provide data over land areas. There are also problems associated with the different radiative behaviour of dust particles of different sizes (coarse dust particles work best, fine dust is more problematic), and there are also problems associated with confusion with aerosols produced by biomass burning. Despite these limitations, the IDDI provides a useful semi-quantitative estimate of atmospheric dust loadings over the Sahara (Brooks and Legrand, 2000).

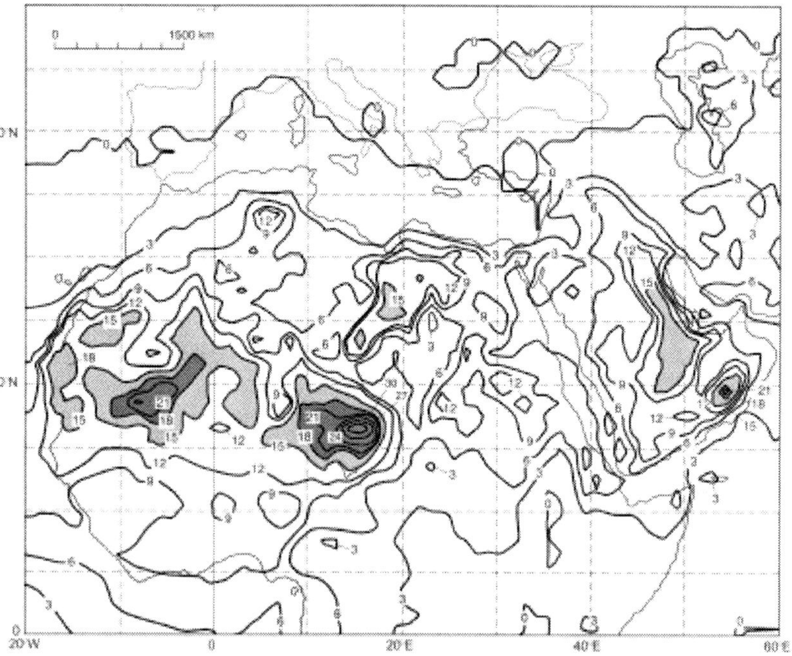

Fig. 3: Annual mean Aerosol Index, highlighting the two main Saharan dust sources, derived from TOMS. Note also the significant sources in the Arabian peninsula (Goudie and Middleton 2001)

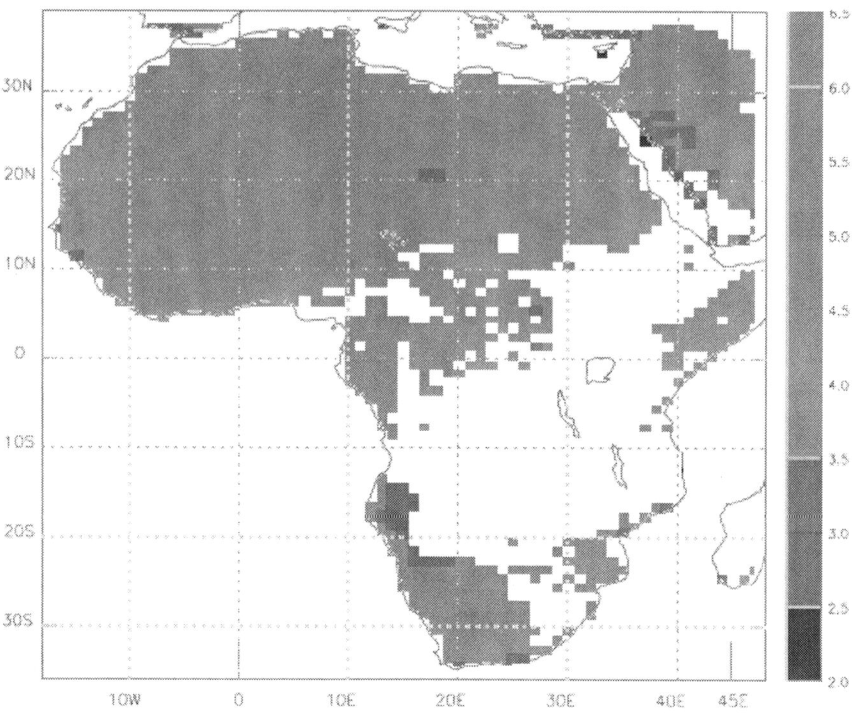

Fig. 4: Long-term average annual IDDI values 1984–1993 (Brooks 1999)

Another important remote sensing tool for monitoring dust is MODIS, the Moderate Resolution Imaging Spectroradiometer, carried on board in both the Terra and Aqua missions. MODIS has 36 spectral bands and operates with three spatial resolutions (250, 500, and 1,000 m). These resolutions are much higher than the 1 degree grids of TOMS and IDDI images. As a result, whereas TOMS and IDDI records can only identify regions of high dust transport, MODIS images can identify the specific landforms generating the dust. These are often ephemeral lake surfaces, where winds remove fine-grained lacustrine sediments deposited during periodic inundations (Bryant et al., 2007). MODIS has a very wide swath (2,330 km) similar to that of the Advanced Very High Resolution Radiometer (AVHRR) carried on board the NOAA series of satellites. The MODIS Rapid Response System was developed to provide daily satellite images of the Earth in near real time, primarily for the purpose of monitoring wildfires and biomass burning. True- and false-color imagery are available within a few hours of being collected, making the system a valuable resource for monitoring atmospheric dust (Fig. 5), especially when used alongside the Aerosol Robotic Network of sun photometers (AERONET). Images can be downloaded from http://rapidfire.sci.gsfc.nasa.gov/

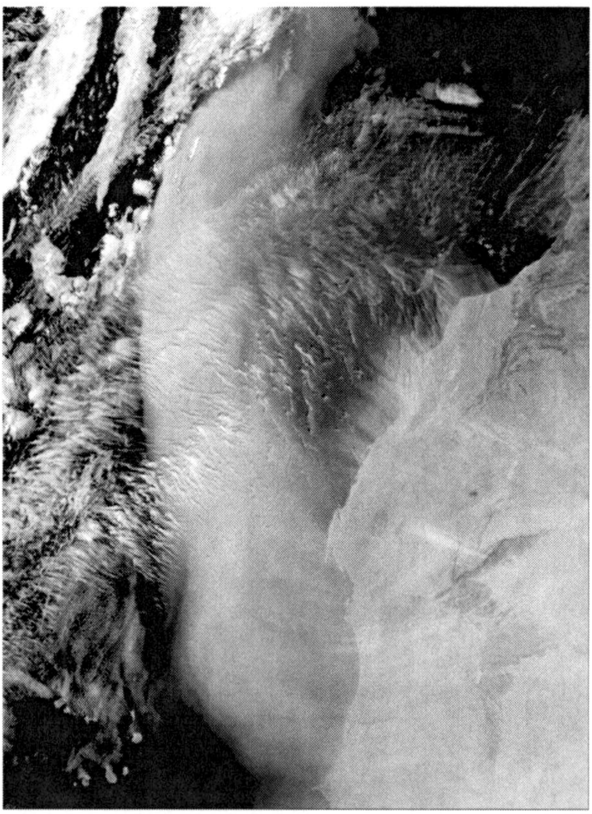

Fig. 5: MODIS Rapid Response image of Saharan dust spreading northward over the Canary Islands, March 4, 2004 (Jeff Schmaltz, MODIS Rapid Response Team, NASA/GSFC)

3 New insights on dust production from remote sensing

Remote sensing res-earch has demonstrated that the global major dust sources are in dryland endoreic basins. Many of these are periodically inundated following episodic heavy rainfall events (Bryant et al., 2007), but many of the basins held much larger lakes during humid climate intervals in the Quaternary (Armitage et al., 2007), and it is this material which is eroding to produce the dust (Washington et al., 2006). These Quaternary palaeolake sediments tend to be fine grained, poorly indurated and easily deflated, often leaving evidence in the form of yardangs (Goudie, 1999). The distribution of these palaeolake sediments can be mapped by remote sensing (White and Eckardt, 2006).

Remote sensing has also highlighted the dustiest place on Earth, the Bodélé Depression in Tchad (Washington and Todd, 2005). This was once part of Lake MegaTchad (Drake and Bristow, 2006), and consists of soft diatomite sediments, easily deflated by aeolian erosion. Recent work undertaken as part of the Bodex experiment (2005) has revealed that these friable sediments are affected by a strong erosive wind called the Bodélé low level jet, which coincides with the exit gap of North-eastern lies between the Tibesti and Ennerdi mountains (Washington and Todd, 2005). Another characteristic which may help to explain the dusty nature of Bodélé is the presence of aeolian sand transport across the diatomite surfaces, to provide a "tool" to kick up the dust. Dune fields are considered to be potential dust sources, by abrasion of the clay coatings on quartz grains during saltation (Bullard and White, 2005), but where dunes move over softer palaeolake sediments, the potential for erosion is much greater. Quartz barchans can be seen to change rapidly to diatomite when they move out onto the diatomite palaeolake surfaces of Bodélé, in Landsat images (Fig. 6). The Bodex field experiment found that the most prolific mechanism of dust production is the breakdown of saltating diatomite flakes as they collide with each other and with the surface. Large barchans act as 'dust mills' causing disturbance of airflow, particularly turbulence and convergence on their "horns". Thus dust production is localized within the main exposure of diatomite to places where there are the most dunes, as on the eastern and northern fringes of the exposure (Warren et al., 2007).

Dust deposition is an inevitable consequence of the dust cycle in drylands. Most deposition occurs in the peri-desert regions, where increasing rainfall washes it from the atmosphere, forming significant deposits of loess, such as are found on the Matmata plateau in southern Tunisia, and along the Djebel Nafusa in Libya, where it forms a useful material from which to carve subterranean dwellings. The distribution of loess can also be mapped by remote sensing using mineral mapping techniques; in the Matmata region, loess was found both as valley and basin fill on the Matmata plateau and as loess-derived soils and stone pavement surfaces on the piedmont slopes surrounding the plateau (White et al., 2002).

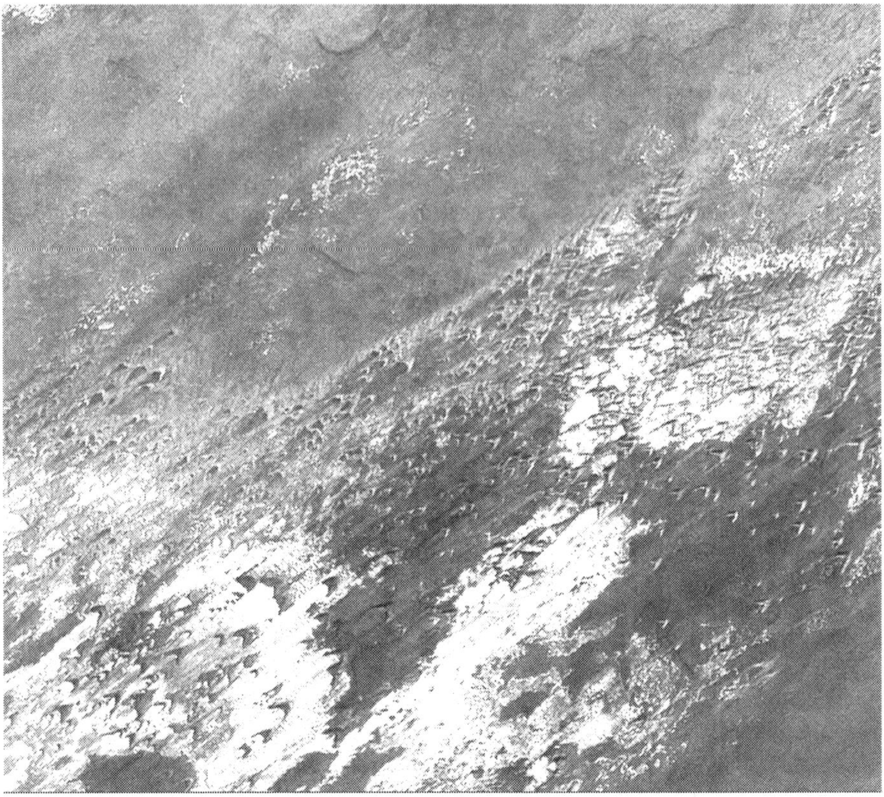

Fig. 6: Landsat Thematic Mapper Image of the Bodélé Depression, quartz barchans in the northwest of the image, transforming rapidly to diatomite barchans as they move in a southwest direction onto the palaeolake surface (Global Land Cover Facility, University of Maryland)

4 Conclusions

Remote sensing is an important tool for monitoring the desert dust cycle. Three main remote sensing methodologies have been used; the TOMS Aerosol Index, the Meterosat IDDI, and MODIS Rapid Response images. These data have provided new insights into the dust cycle, identifying the major sources as predominantly endoreic drainage basins, many of which were occupied by lakes during more humid climate intervals in the Late Quaternary. The residual palaeolake deposits have a fine particle size distribution and are easily entrained by aeolian erosion. Turbulent winds then carry this material high into the atmospheric column, where it can be exported long distances, such as across the Atlantic Ocean (Prospero et al., 1983).

Bibliography

Armitage S.J., Drake N.A., Stokes S., El-hawat A., Salem M.J., White K., Turner P., McLaren, S.J. 2007. *Multiple phases of North African humidity recorded in lacustrine sediments from the Fazzan basin, Libyna Sahara*, in Quaternary Geochronology, n.**2**, 181–186.

Brooks, N. 1999. *Dust-climate interactions in the Sahel-Sahara zone of northern Africa, with particular reference to late twentieth century Sahelian drought.* Unpublished Ph.D. Thesis, University of East Anglia.

Brooks N., Legrand, M. 2000. *Dust variability and rainfall in the Sahel*, in McLaren S. and Kniveton D. (Eds.) Linking Climate Change to Land-Surface Change. Kluwer, Dordrecht, 1–25.

Bryant R.G., Gigg G.R., Mahowald N.M., Eckardt F.D., Ross S.G. 2007. *Dust emission response to climate in southern Africa*, in Journal of Geophysical Research, n.**112**, D09207, doi: 10.1029/2005JD007025.

Bullard, J.E., White, K. 2005. *Dust production and the release of iron oxides resulting from the aeolian abrasion of natural dune sands*, in Earth Surface Processes and Landforms, n.**30**, 95–106, doi: 10.1002/esp.1148.

Buritt B., Hyers A.D. 1981. *Evaluation of Arizona's highway dust warning system*, in Geological Society of America Special Paper, n.**186**, 281–292.

Cakmur R.V., Miller R.L., Torres O. 2004. *Incorporating the effect of small-scale circulations upon dust emission in an atmospheric general circulation model*, in Journal of Geophysical Research, n.**109**, D07201, doi: 10.1029/2003JD004067.

Chiapello, I., Moulin C. 2002. TOMS and METEOSAT *satellite records of the variability of Saharan dust transport over the Atlantic during the last two decades (1979–1997)*, in Geophysical Research Letters, n.**29** (8), 1176, doi: 10.1029/2001GL013767.

Drake N., Bristow C.S. 2006. *Shorelines in the Sahara: geomorphological evidence for an enhanced monsoon from palaeolake Megachad*, Holocene n.**16** (6), 901–911.

Ganor, E., Mamane Y. 1982. *Transport of Saharan dust across the eastern Mediterranean*, in Atmospheric Environment, n.**16**, 581–587.

Goudie A.S. 1999. *Wind erosional landforms: yardangs and pans*, in Goudie, A.S., Livingstone, I., and Stokes, S. (Eds.) Aeolian Environments, Sediments and Landforms. Wiley, Chichester, 167–180.

Goudie A.S., Middleton, N.J. 1992. *The changing frequency of dust storms through time*, in Climatic Change, n.**20**, 197–225.

Goudie A.S., Middleton N.J. 2001. *Saharan dust storms: nature and consequences*, in Earth-Science Reviews, n.**56**, 179–204.

Herman J.R., Bhartia P.K., Torres O., Hsu C., Seftor, C., Celarier E. 1997. *Global distribution of UV-absorbing aerosols from Nimbus 7/TOMS data*, in Journal of Geophysical Research. n.**102**, 16911–16922.

IPCC 2001. *Climate Change 2001.* Contribution of Working Group I to the Third Assessment Report of the Intergovernmental Panel on Climate Change. Cambridge University Press, Cambridge.

Kaufman Y.J., Koren I., Remer L.A., Tanre D., Ginoux P., Fan S. 2005. *Dust transport and deposition observed from the Terra-Moderate Resolution Imaging Spectroradiometer (MODIS) spacecraft over the Atlantic Ocean*, in Journal of Geophysical Research, n.**110**, D10S12, doi: 10.1029/2003JD004436.

Leathers C.R. 1981. *Plant component of desert dust in Arizona and their significance for man*, in Geological Society of America Special Paper, n.**186**, 191–206.

Legrand M., N'Doume C., Jankowiak I. 1994. *Satellite-derived climatology of the Saharan aerosol*, in Lynch D.K. (Ed.) Passive Infrared Remote Sensing of Clouds and the Atmosphere II. Proceedings of SPIE 2309, 127–135.

Legrand M., Plana-Fattori A., N'Doume C. 2001. *Satellite detection of dust using the IR imagery of Meteosat: 1. Infrared difference dust index*, in Journal of Geophysical Research, n.**106** (D16), 18251–18274.

Leon J.F., Legrand M. 2003. *Mineral dust sources in the surroundings of the north Indian Ocean*, in Geophysical Research Letters, n.**30** (6), 1309, doi: 10.1029/2002GL016690.

Leys J. 1999. *Wind erosion on agricultural land*, in Goudie, A.S., Livingstone, I. and Stokes, S. (Eds.) Aeolian Environments, Sediments and Landforms. Wiley, Chichester, 143–166.

Liss P.S., Turner S.M. 2000. *The influence of iron on ocean biology and climate; insights from the IronEx studies in the Equatorial Pacific*, in Hanson R.B., Ducklow H.W. and Field J.G. (Eds.) The Changing Ocean Carbon Cycle, A Midterm Synthesis of Joint Global Ocean Flux Study. Cambridge University Press, Cambridge.

Livingstone I., Warren A. 1996. *Aeolian Geomorphology*. Longman, Harlow.

McFadden L.D., Wells S.G., Jercinovich M.J. 1987. *Influences of aeolian and pedogenic processes on the origin and evolution of desert pavements*, in Geology, n.**15**, 504–508.

McTainsh G. 1999. *Dust transport and deposition*, in Goudie, A.S., Livingstone, I. and Stokes, S. (Eds.) Aeolian Environments, Sediments and Landforms. Wiley, Chichester, 181–211.

Middleton N.J. 1985. *Effect of drought on dust production in the Sahel*, in Nature, n.**316** (6027), 431–435.

Middleton N.J. 1986. *Dust storms in the Middle East*, Journal of Arid Environments, n.**10**, 83–96.

Middleton N.J., Goudie A.S. 2001. *Saharan dust, sources and trajectories*, in Transactions of the Institute of British Geographers, n.**26** (2), 165–181.

Middleton N.J., Goudie A.S., Wells G.L. 1986. *The frequency and source areas of dust storms*, in Nickling, W.G. (Ed.) Aeolian Geomorphology. Allen & Unwin, Boston, pp. 237–260.

Prospero J.M., Glaccum R.A., Nees R.T. 1983. *Atmospheric transport of soil dust from Africa to South America*, in Nature, n.**320**, 570–572.

Torres O., Bhartia P.K., Herman J.R., Sinyuk A., Holben B. 2002. *A long term record of aerosol optical thickness from TOMS observations and comparison to AERONET measurements*, in Journal of Atmospheric Science, n.**59**, 398–413.

Warren A., Chappell A., Todd M.C., Bristow C.S., Drake N.A., Engelstaedter S., Martins V., M'bainayel S., Washington R. 2007. *Dust-raising in the dustiest place on earth*, in Geomorphology, n.**92**, 25–37, doi:10.1016/j.geomorph.2007.02.007.

Washington R., Todd M.C. 2005. *Atmospheric controls on mineral dust emission from the Bodélé Depression, Chad: the role of the low level jet*, in Geophysical Research Letters, n.**32**, L17701, doi: 10.1029/2005GL023597

Washington R., Todd, M. Middleton, N.J., Goudie, A.S. 2003. *Dust-storm source areas determined by the total ozone monitoring spectrometer and surface observations*, in Annals of the Association of American Geographers, n.**93** (2), 297–313.

Washington R., Todd, M.C., Engelstaedter, S., Mbainayel, S., Mitchell F. 2006a. *Dust and the low-level circulation over the Bodélé Depression, Chad: observations from BoDEx 2005*, Journal of Geophysical Research, n.**111**, D03201, doi: 10.1029/2005JD006502.

Washington R., Todd M.C., Lizcano G., Tegen I., Flamant C., Koren, I., Ginoux P., Engelstaedter S., Bristow C.S., Zender C.S., Goudie A.S., Warren A., Prospero J.M. 2006b. *Links between topography, wind, deflation, lakes and dust: the case of the Bodélé Depression, Chad*, in Geophysical Research Letters, n.**33**, L09401, doi: 10.1029/2006GL025827.

Wasson R.J., Nanninga P.M. 1986. *Estimating wind transport of sand on vegetated surfaces*, in Earth Surface Processes and Landforms, n.**11**, 505–514.

Wheaton E.E. 1992. *Prairie dust storms, a neglected hazard*, in Natural Hazards, n.**5**, 53–63.

White K., Eckardt F.D. 2006. *Geochemical mapping of carbonate sediments in the Makgadikgadi basin, Botswana, using moderate resolution remote sensing data*, in Earth Surface Processes and Landforms, n.**31**, 665–681, doi: 10.1002/esp.1289.

White K., Livingstone I., Gurney S., Dearing J., Bateman M. 2002. *Post-processing of mineral mixture maps for mapping surficial materials: the example of the Matmata loess, southern Tunisia*, in International Journal of Remote Sensing, n.**23**, 3091–3106, doi: 10.1080/01431160110104674

Worster D. 1979. Dust Bowl; the Southern High Plains in the 1930s. Oxford University Press, Oxford.

SPACE FOR THE UNCCD AND THE DESERTWATCH PROJECT

GAETANO PACE[1], GAIA VAGLIO LAURIN,[2] LUCIO PIRES DO ROSARIO,[3] AND MAURIZIO SCIORTINO[4]

[1]*Advanced Computer Systems, via della Bufalotta, 378, 00139, Rome, Italy g.pace@acsys.it*
[2]*Advanced Computer Systems, via della Bufalotta, 378, 00139, Rome, Italy g.laurin@acsys.it*
[3]*Direcção-Geral Dos Recursos Florestais, Ministry of Agriculture, Avenida João Crisóstomo, 26-28 1069-040 Lisbon, Portugal*
lucio.rosario@dgrf.min-agricultura.pt
[4]*National Committee for Combating Drought and Desertification, Via Cristoforo Colombo, 112, 00147, Rome, Italy*
maurizio.sciortino@casaccia.enea.it

Abstract

The ESA DesertWatch project goal was to develop an Information System for assessing and monitoring desertification using EO technologies, in order to support local authorities in reporting to the UNCCD. The project has selected the most robust and reliable methodologies developed during the related research projects, integrating them into a unique environment, which is easy to use, as automatic as possible, and cost effective. The system outputs consist in a series of desertification indicators and products, generated at regional, national and sub-national scale. For demonstration purposes these products have been implemented over large areas of Portugal, Italy and Turkey for the years 1984, 1994 and 2004. The validation phase has shown that the results are in line with the user requirements. The system has been much appreciated by the final users and will physically be utilized by various national and local authorities. The possibility of expanding the system in extra-European countries is under evaluation.

Keywords: Desertification, Earth Observation, monitoring, Landsat, MERIS, ESA

1 Introduction

Desertification is defined by the United Nations Convention to Combat Desertification (UNCCD) as "land degradation in the arid, semi-arid and dry sub-humid areas resulting from various factors, including climatic variations and human activities" (UNCCD, 1994). In Northern Mediterranean 99.4 million hectares show signs of desertification, corresponding to about 32% of territory with arid, semi-arid and dry sub-humid climate (UNCCD, 2000). As part of the Data User Element (DUE) of the Earth Observation Envelope Programme, the European Space Agency (ESA) launched in the 2004 the DesertWatch (DW) project, whose requirements have been defined in collaboration with North Mediterranean Region National Focal Points of

A. Marini and M. Talbi (eds.), *Desertification and Risk Analysis Using High and Medium Resolution Satellite Data*,

the UNCCD. DW exploits the potential of Earth Observation (EO) technology for monitoring desertification, building an Information System (IS) aimed at supporting the UNCCD and the parties in the convention implementation.

2 Objectives

The main project objective is to monitor desertification at various scales and through time, developing an IS which can be operationally adopted by national and local authorities for reporting to the UNCCD and for local rules and policy development. The system shall be easy to use, as automatic as possible and, most importantly, minimise the need of costly input data. Many past and ongoing national and European research projects, such as MODULUS, MEDALUS, LADAMER, GeoRange, DeSurvey and RIADE combined EO and field data for monitoring land degradation at regional level. The DW IS is based on those research results, in particular on the commonly accepted MEDALUS approach, which integrates biophysical, socio-economic and management parameters into a holistic view, to identify Environmental Sensitive Areas to desertification. The DW IS adapted this MEDALUS approach to compose a mainly EO-based system, which shall be a standardised, robust, reporting-oriented, and user-friendly tool.

3 Study areas

The DW IS shall be able to generate a set of desertification related indicators and products over large areas. For demonstration purposes the project has identified a number of study areas in Italy, Portugal and Turkey. In particular, the following areas have been analysed:

- Portugal: whole continental country
- Italy: Sardinia, Basilicata and Sicily regions
- Turkey: Konia Karapinar and Corouh watershed

Multitemporal mapping for the years 1984, 1994 and 2004 has been conducted over the study areas to demonstrate the system capability of monitoring changes and trends.

4 Methodology

Monitoring desertification extent requires the evaluation of a complex set of indicators, related to climatic (e.g.: rainfall, evapotranspiration, aridity indexes), bio-physical (e.g.: morphology, soil and vegetation properties) socio-economic (e.g.: population density and age, employment) and management (e.g. policies, protected areas, master plans) factors (Enne and Zucca, 2000). The mentioned MEDALUS project approach, identifies Environmental Sensitive Areas (ESAs) through the definition of climate, vegetation, soil and management quality indexes (Kosmas et al., 1999).

Some of these indexes can be successfully assessed exploiting remote sensing technologies and geomatic applications (Boer and Puigdefabregas, 2005; Hill and Peter, 1996; Hill et al., 2006). The DW IS uses primarily EO data, in combination with some ancillary data, into a seamless data processing facility.

To assess the needed indexes the following principal approaches/techniques were used:

- Land cover mapping
- Spectral mixture analysis
- Land degradation index assessment
- Spatial dynamic modelling

These techniques will be briefly described in the following paragraphs, along with the corresponding indicators/products generated.

4.1 Land cover mapping

Land cover maps are invaluable instruments for environmental monitoring and territorial planning. In desertification monitoring, a land cover map provides a number of indicators which can be directly extracted from the map, thus providing useful and easily accessible information.

Even if land cover maps are usually available for the EU area (i.e. CORINE 2000 project (Büttner et al., 2004)), those maps are rarely updated or organized at the needed thematic detail for desertification monitoring purposes.

Hence, the first concern of DW project was to provide users with updated, cost-effective and accurate land cover maps, so that frequent monitoring and trends detection could also take place.

From those maps, the indicators needed to identify the ESAs areas are thus extracted by the DW IS facility.

Other indicators, not used in the MEDALUS approach are also identified, as requested by users and UNCCD National Focal Points. All indicators can be spatially aggregated, providing statistics according to administrative boundaries (municipal, district, regional, national level etc.).

Land cover mapping relies on a fully automatic procedure that uses as input:

- Two pre-processed Landsat satellite images (possibly from spring and fall seasons)
- Digital Elevation Model, Slope and Aspect
- CORINE land cover map (CLC)

All the steps are conducted into the DW IS.

Pre-processing includes georeferentiation orthorectification and radiometric correction. The most significant 5 bands are selected out of the 2 images input, using PCA technique and are merged with altitude, slope and aspect maps to produce an 8-band file ready to be segmented. A segmentation procedure, based on a multi-scale hierarchical weighted aggregation approach (Galli and De Candia, 2005) is applied, allowing the performance of an object oriented classification.

The training set can be extracted either manually (e.g. furnished from an external source) or automatically, via a filtering procedure based on the pre-existing CLC map.

The image classification adopts the Maximum Likelihood algorithm and is designed at producing three maps: one obtained using the bands for each season and one using the joined bands of both seasons. Burnt areas and internal and marine waters are masked with CLC dataset. A weighted consensus procedure, which uses the three classification results plus the original CLC map, is used for creating a unique land cover map. This map is cleaned up with a clump and sieve procedure, which creates the final map having a Minimum Mapping Unit (MMU) of 5 ha.

The adopted legend is an aggregation of the CLC legend into 12 classes (see Table 1), specifically identified to provide the necessary information for desertification assessment purposes.

Table 1: Land Cover legend used in DW

Coniferous forest
Deciduous forest
Mixed forest and Mediterranean maquis
Grassland
Permanent trees and heterogeneous
Agriculture
Not irrigated crops
Irrigate land
Artificial areas
Wetlands
Water
Bare land
Forest fires

From the land cover map a number of indicators are assessed by direct land cover class extraction:

- Forested areas (simply aggregating the vegetation classes of interest, i.e. Deciduous Forest, Coniferous Forest)
- Soil sealing (equal to the Artificial areas class)
- Irrigated lands (equal to the Irrigated area class)

Other indicators are assessed through specific spatial analyses:

- Forest Fragmentation (defined as in (Hurd et al., 2002)): estimates forest quality over an area which is usually coincident with an administrative boundary, in terms of forest density (amount of forested area over total area) and continuity (number of patches/adjacent pixels into the forested area).
- Re-naturalized Areas: estimates the land cover change from agriculture to natural vegetation classes and relates it with demographic changes. Where a change from agriculture to natural vegetation occurs in a given area simultaneously with a population reduction (estimated by ancillary data such as census), the area undergoes a re-naturalization.

- Burnt Areas: a specific Fire Tool has been developed and integrated into the DW IS to estimate those areas. The tool uses the Normalised Burnt Ratio (based on the analysis of Landsat bands 4 and 7) to detect areas interested by fires in the time period comprised by the two seasonal satellite images acquisition dates.

4.2 Spectral mixture analysis

The semi-empirical model, known as Spectral Mixture Analysis (SMA), describes spectral reflectance signatures as a mixture of few prototype spectra, also called endmembers (Adams et al., 1989; Smith et al., 1990). In natural environments, the principal endmembers are typically represented by vegetation, soil, bedrock and shade, which commonly mix at the sub-pixel scale producing mixed-pixel spectra. SMA attempts to unmix such multispectral reflectance in order to assess, for each pixel, the relative abundance of each component. Using this technique it is possible to retrieve the following useful indicators (Hill, 1993; Baret and Jacquemoud, 1994; Hill and Schütt, 2000):

- Vegetation Abundance (VEA)
- Normalized Soil/Rock Ratio (NSRR)

The vegetation abundance, which simply corresponds to the proportion of the vegetation endmember, besides being an indicator in itself, can be used in conjunction with other parameters for the assessment of the vegetation quality index, as explained in section 5.5. In Mediterranean ecosystems, NSRR can be seen as a surrogate of soil erosion, especially if the trend analysis shows a decrease of the ratio as explained in (Hill and Schütt, 2000). This technique can be used with both Landsat and MERIS EO data, for obtaining maps at local and national scales.

4.3 Land degradation index

The Land Degradation Index (LDI), whose computational details are described in Hill and Peter, 1996, exploits remote sensing images, in conjunction with climatic and physiographic parameters, for assessing the landscape status with respect to its natural resources potential. It works on the assumption that sites in good condition will show outputs in their local water balance mostly through evapotranspiration rather than through runoff.

Vegetation abundance, computed as indicated in the previous section, is used to estimate vegetation status. Input data fall into the following categories:

- Remote sensing data: 2-years time series, with at least six frames per year, for VEA calculation
- Climatic data: rainfall and potential evapotranspiration
- Topographic data: obtained from DEM
- Daily rainfall series from a single location
- External physiographic layers: lithology, soil types, terrain types, land cover

Based on the external physiographic layers, homogeneous land systems are identified and their status is expressed in terms of number of standard deviations from the population mean. The final values are then converted into three broad classes, that indicate poor, normal and good land conditions. This index can be computed at various resolutions, according to the required mapping scale. For this project both Landsat and MERIS data have been used.

4.4 Spatial dynamic modelling

A spatial modelling tool has been developed for simulating Scenarios of Desertification (named ScenDes). The tool is able to generate possible scenarios for desertification monitoring, starting with a land use map at a certain time and simulating the land use evolution during the years. The complex simulation model uses local, zonal and neighbouring map-based rules for computing the land use transition potential. For each map pixel, the system computes the probability to change (transition potential) into each of the possible land use classes. In addition to this, based on socio-economic data, such as population, growth rate, job availability, etc., the ScenDes evaluates the total demand for each land use. For each simulation step (in these simulations the time step is set to 1 year) the system will try to meet the overall demand for each land use class, starting to change the pixels with highest transition potential, until the total demand is met. The core algorithm, based on the cellular automata paradigm, is described in detail in (Batty and Xie, 1994; Couclelis, 1985; White and Engelen, 1993; White and Engelen, 1997; White et al., 2004). The major input variables are listed hereafter:

- Land use maps corresponding to at least two time periods, required for setting the initial status and for calibration purposes.
- Physiographic data (e.g.: soil type, digital terrain model, geological maps, climatic layers), used for defining the transition potentials, based on local suitability.
- Communication network (e.g.: roads, railways, with location of access points, stations), required for computing the transition potentials based on accessibility.
- Zoning maps (master plans, protected areas, natural reserves, etc.), required for identifying areas where certain land use types are not allowed in the simulation (e.g. the urban land use is not allowed in a natural reserve).

The simulation is typically carried out on periods of 10–20 years and allows visualising possible land use scenarios. These land uses are finally translated into desertification indicators, in a way similar to what indicated in section 4.1.

4.5 Validation approach

The collection of validation data has been carried out by local teams identified by the DW end users, under the coordination of the project team. Validation data have been acquired directly on the field and, to a lesser extent, using photointerpretation. Different validation approaches have been used for the various product types described above.

For whole validation purposes 3 areas have been identified in Italy (Fig. 1, left) and 10 areas in Portugal (Fig. 1, right), covering a total of about 25,000 km^2. To validate the land cover maps and related desertification indicators, sampling points were selected using a stratified random sampling method, according to land cover type abundance.

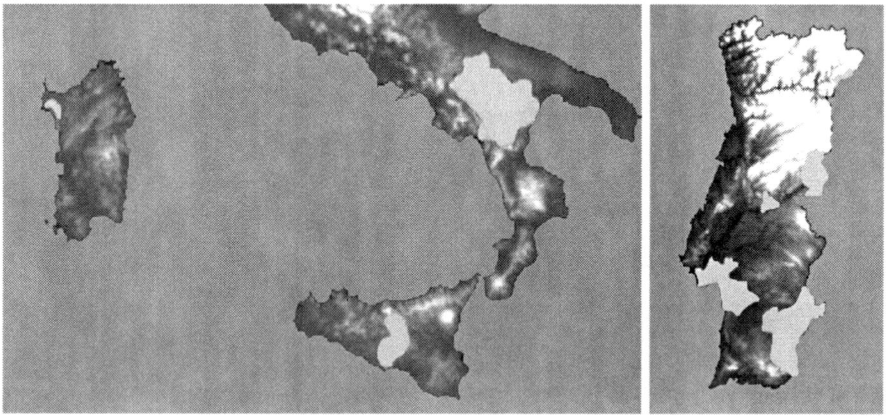

Fig. 1: Validation areas in Italy (left) and Portugal (right)

The validation of SMA products was based on the comparison of the relative abundances of vegetation, soil and exposed rock against the field measurements. For validating the LDI product, a number of qualitative attributes were estimated on the field, corresponding to ground attributes (topography, rock type, landform, etc.) vegetation attributes (structure, density, condition, vigour, etc.), soil degradation features (signs of erosion, rock fragments, exposed deep horizons, etc.). The final LDI validation shall be based on expert judgment of the produced maps, versus the field observation.

The ScenDes product can be validated using at least 3 land cover maps acquired at different times. The 1st and 2nd land cover map are used for the calibration of the system, where the 3rd map shall be computed against the simulated scenarios.

5 Results

5.1 Land cover related products

An example of land cover map generated over the whole Portugal is reported in Fig. 2. Land cover maps have been generated and validated against field data over Italy and Portugal, with an accuracy ranging between 75% and 85%. For Turkey the validation has been conducted against land cover maps provided by local partners, obtaining similar values of accuracy.

The indexes derived from the land cover map are not directly validated, considering that the validation of the original land cover map is most appropriate. Anyhow, the accuracy of the indexes, often being aggregation of more than one land cover type, ranges usually over 80%.

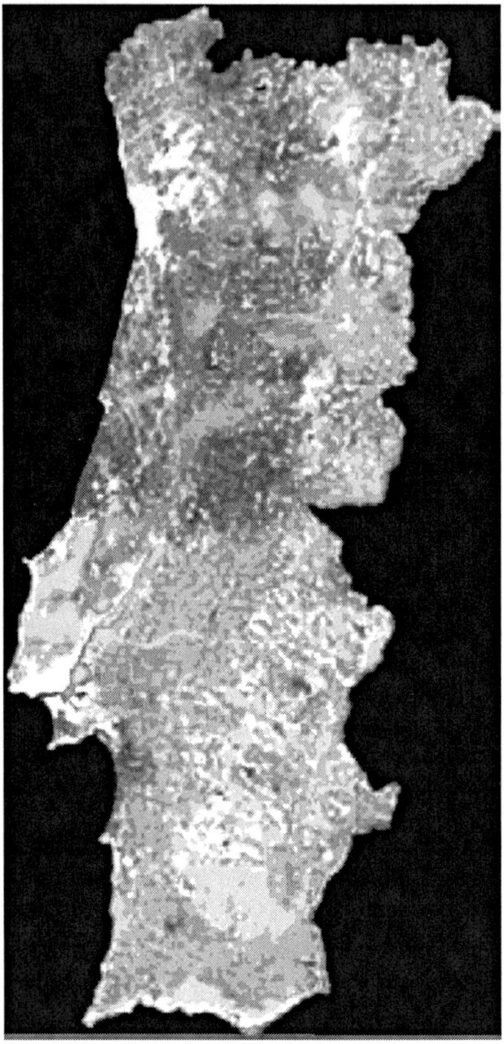

Fig. 2: Land Cover map produced over Portugal

5.2 Spectral mixture related products

The VEA and NSRR maps have been generated over all the project demonstration areas. A preliminary validation analysis has been carried out in Sicily, using 45 sampling plots and the Landsat-based VEA map. The scatterogram of the ground based estimates vs. the VEA best pixel within 30 m, reported in Fig. 3, shows a good correlation, with

$R2 = 0.74$. Validation of NSRR is still in progress.

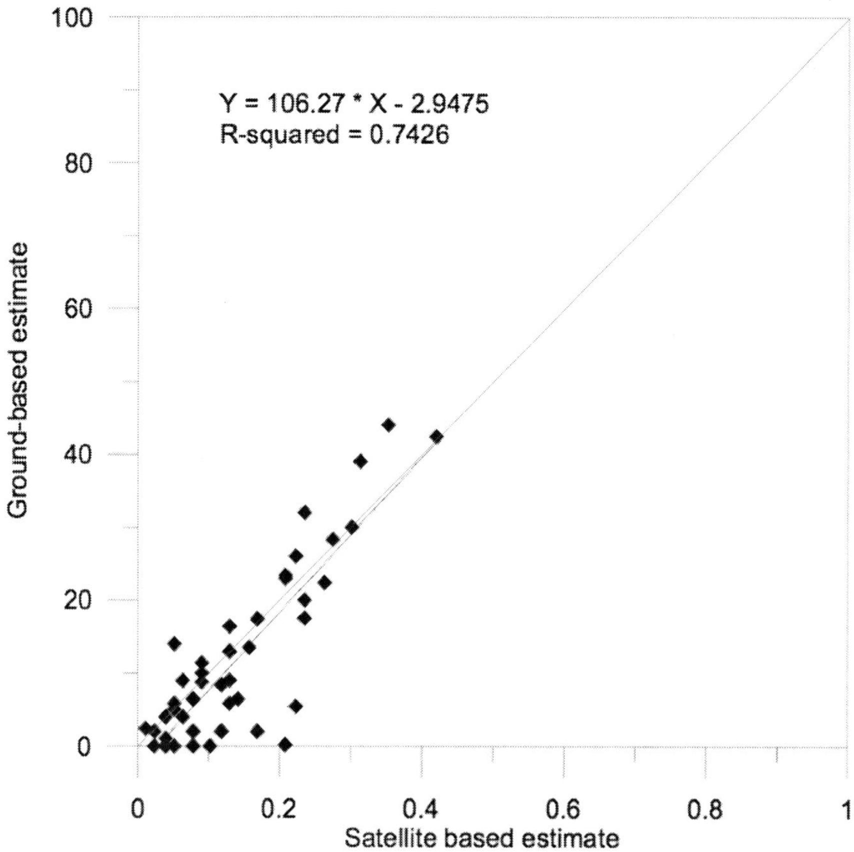

Fig. 3: Scatterogram of MERIS vs. Landsat derived vegetation abundances

5.3 Land Degradation Index products

The LDI maps have been generated in most project areas. An example of map generated in the Southern part of Portugal is represented in Fig. 4. The product validation is still in progress.

5.4 Desertification scenarios

The ScenDes tool for the generation of desertification scenarios has been set-up and calibrated over 4 areas, corresponding to the whole continental Portugal, and the Italian regions of Sicily, Sardinia and Basilicata. The results have been judged and discussed with expert panels during ad-hoc thematic workshops.

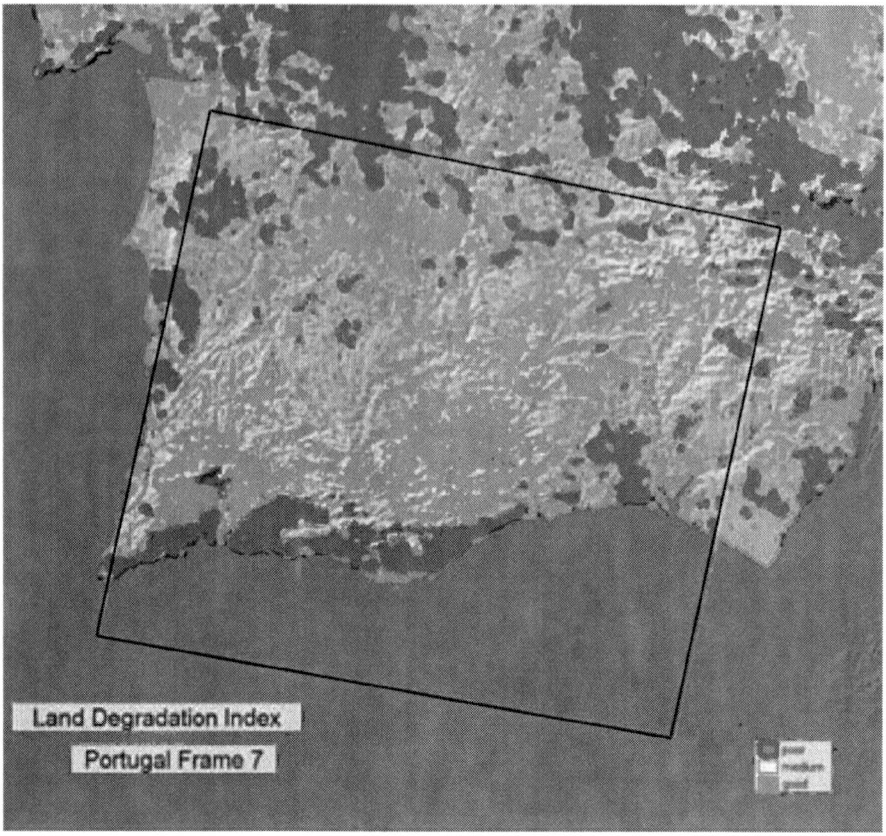

Fig. 4: LDI map generated over a South Portugal area

5.5 Generation of the Environmental Sensitive Areas (ESAs) map

According to the previously mentioned MEDALUS methodology, it is possible to generate the Environmental Sensitive Areas (ESA) map, which gives the susceptibility to desertification. The ESA map is computed according to the following formula:

$$\mathrm{ESA} = (\mathrm{CQI} * \mathrm{VQI} * \mathrm{SQI} * \mathrm{MQI})\ ¼ \tag{1}$$

where: CQI is the climate quality index; SQI is the soil quality index; VQI is the vegetation quality index; MQI is the management quality index. CQI and SQI are computed using external GIS data. The VQI is defined as follows in (Kosmas et al., 1999):

$$\mathrm{VQI} = (\mathrm{FR} * \mathrm{EP} * \mathrm{DR} * \mathrm{PC})\ 1/4 \tag{2}$$

where: FR is the fire risk; EP is the erosion protection; DR is the drought resistance; PC is the plant cover. The DW land cover maps are used for assessing the FR, EP and DR parameters: a different value is assigned to each land cover class according to the model described in (Kosmas et al., 1999), with some slight modifications needed to use EO data as main input source. The PC parameter is simply derived from the VEA products obtained with SMA technique. The MQI is defined as follows in (Kosmas et al., 1999):

$$MQI = (LUI * PP) 1/2 \tag{3}$$

where: LUI is the land use intensity; PP indicates the protection policies.

The LUI is calculated on the basis of the DW land cover, using a weighting procedure similar to that described above. Concerning PP, calculation, information concerning administrative limits of protected areas (National Parks, nature reserves, etc.) should be provided by the users. Thereafter, a weighting system assigns different score to areas covered by natural or semi-natural vegetation in DW land cover map, which are protected through different policies and rules.

6 Conclusions

The obtained results are in line with requirements defined by the project. Goodness of input data, including ancillary ones, obviously influences the final results: the flexibility of the DW IS permits to update the products in terms of frequency, producing new information at low cost, and in terms of quality, using better ancillary data as they become available. The validation phase has been a chance to evaluate the DW IS and the products with an active end user involvement. The registered level of satisfaction for the system and the indicators and maps produced has been very high. The DW IS will be soon installed in the appropriate head offices (Ministries, local territorial authorities, etc.) and routinely used to report desertification trends. At present, an extension of the DW project to extra European countries is under evaluation, to adapt the system to different data and environments, with a special focus on developing countries where the majority of land desertification occurs.

Acknowledgements

The authors wish to thank the whole DesertWatch team, which includes: University of Sassari, Italy, Nucleo di Ricerca sulla Desertificazione (NRD) and Ente per le Nuove Tecnologie, l'Energia e l'Ambiente (ENEA), Italy for the support to the generation of the land cover map related indicators, and to the validation campaign; University of Trier, Remote Sensing Department (RSD), for the development of the SMA methodology and related products; Estación Experimental de Zonas Áridas (EEZA) from Consejo Superior de Investigaciones Científicas (CSIC), Spain, for the development of the LDI methodology and related products; Research Institute for Knowledge Systems (RIKS) for the development, implementation and calibration of the ScenDes tool and of the related desertification indicators.

Bibliography

Adams J.B., Smith M.O., and Gillespie A.R. 1989. *Simple models for complex natural surfaces: a strategy for the hyperspectral era of remote sensing*. Proc. IGARSS '89 Symposium, July 10–14, Vancouver, Canada, 16–21.

Baret F. and Jacquemoud S. 1994. *Modelling canopy spectral properties to retrieve biophysical and biochemical characteristics*, in Hill J. and Mégier J. (eds.), 1994. Imagine Spectrometry – A Tool for Environmental Observations. Euro Courses Remote Sensing, n. **4**, pp. 145–168.

Batty M. and Xie Y. 1994. *From cells to cities*, in Environment and Planning, n.**B 21**, pp. 31–48.

Boer M.M. and Puigdefabregas J. 2005. *Assessment of dryland condition using spatial anomalies of vegetation index values*, in International Journal of Remote Sensing, 2005, VOL 26, n. **18**, pp. 4045–4066.

Büttner G., Feranec J., Jaffrain G., Mari L., Maucha G., and Soukup T. 2004. *The CORINE Land Cover 2000 Project*, in European Association of Remote Sensing Laboratories Proceedings.

Couclelis H. 1985. *Cellular worlds: a framework for modelling micro-macro dynamics*, in Environment and Planning, n.**B 17**, pp. 585–596.

Enne G. and Zucca C. 2000. *Desertification indicators for the European Mediterranean region. State of the art and possible methodological approaches*. ANPA, Rome, 121.

Galli L. and De Candia D. 2005. *Multispectral Image Segmentation via Multiscale Weighted Aggregation Method, submitted and accepted for SPIE Remote Sensing 2005*, 12th International Symposium, September 2005.

Hill J. 1993. *Monitoring land degradation and soil erosion in Mediterranean environments*, in ITC Journal, 1993, pp. 323–331.

Hill J. and Peter D. (eds.). 1996. *The use of remote sensing for land degradation and desertification monitoring in the Mediterranean Basin. State of the art and future research.* Proc. of a Workshop, jointly organized by JRC/IRSA and DGXII/D-2/D-4, Valencia, 13–15 June 1994, Valencia, EUR 16732 EN (Office for Official Publications of the European Communities: Luxembourg).

Hill J. and Schütt B. 2000. *Mapping complex patterns of erosion and stability, in Dry Mediterranean ecosystems*, in Remote Sensing of Environment, n.**74**, pp. 557–569.

Hill J., Jarmer T., Udelhoven T., and Stellmes M. 2006. *Remote sensing and geomatics applications for desertification and land degradation monitoring and assessment*, in Escadafal R. and Paracchini M.L. (eds.), Geometrics for Land and Water Management: Achievements and Challenges in the Euromed Context, International Workshop, Joint Research Centre, Italy, 23–25 June 2004, 15–22.

Hurd J.D., Wilson E.H., and Civco D.L. 2002. *Development of a forest fragmentation index to quantify the rate of forest change*. Proc. 2002 ASPRS Annual Convention, Washington, DC 10 p.

Kosmas C., Kirkby M., and Geeson N. 1999. *Manual on key indicators of desertification and mapping environmentally sensitive areas to desertification*, in the MEDALUS project – Mediterranean desertification and land use. European Commission, Brussels.

Smith M.O., Adams J.B., and Gillespie A.R. 1990. *Reference endmembers for spectral mixture analysis*. Proc. 5th Australasian Remote Sensing Conference, Perth, pp. 331–340.

United Nation Convention for Combating Desertification (UNCCD). 1994. *The convention*, Part 1, Introduction.

United Nation Convention for Combating Desertification (UNCCD). 2000. *ICCD/COP(4)/3/Add.3 (A)*.

White R. and Engelen G. 1993. *Cellular automata and fractal urban form: a cellular modelling approach to the evolution of urban land use patterns*, in Environment and Planning, n.**A 25**, pp. 1175–1199.

White R. and Engelen G. 1997. *Cellular automata as the basis of integrated dynamic regional modelling*, in Environment and Planning, n.**B 24**, pp. 235–246.

White R., Straatman B., and Engelen G. 2004. *Planning scenario visualization and assessment, a cellular automata based integrated spatial decision support system,* in Goodchild M.F. and Janelle D.G. (eds.), Spatially Integrated Social Science, Oxford University Press, New York, 420–442.

THE NEW THEMATIC STRATEGY FOR PROTECTING SOILS IN EUROPE

GIOSUÈ LOJ
Member of Italian Soil Observatory
ERSAT – Tecnica in Agriculture, Cagliari, Italy gloj@tiscali.it

Abstract

Soil is a vital and largely non-renewable resource which is increasingly under pressure. Many EU policies (water, waste, chemicals, industrial pollution prevention, nature protection, pesticides, agriculture) are contributing to soil protection. But as these policies have other aims and other scopes of action, they are not sufficient to ensure an adequate level of protection for all soil in Europe. For this reason, the Commission adopted a Soil Thematic Strategy (COM (2006), 231) and a proposal for a Soil Framework Directive (COM (2006), 232) with the objective to protect soils across the EU.

The Soil Thematic Strategy outlines major soil functions: food and other biomass production, storing, filtering, and transformation, habitat and gene pool, physical and cultural environment for mankind, source of raw materials, and points out main soil threats such as erosion, salinization, organic matter decline, compaction, and landslides.

Keywords: Soils, desertification, thematic strategy

1 Introduction

Erosion is a natural process enhanced by human activities. Soil erosion is driven by both natural and anthropogenic causes. The latter increase the magnitude and frequency of the process.

In Europe the agriculture is one of the main drivers of unnatural soil erosion, since many farming practices are soil-unfriendly and almost half of the European territory is intensively farmed, often in an unsustainable way. Farmer's management decisions are determined by market conditions, technological development, changes in the wider economy, particularly the rising of relative cost of labour, and a range of structural changes. As a result, pressures on the environment increased by changes in land structure (land levelling or disappearing of landscape elements such as hedges, shelterbelts, etc.), changes in crop patterns and inappropriate agricultural practices.

In the more disadvantaged areas, land abandonment is the main driver, potentially leading to desertification if the soil productivity and erosion tolerance are low, and the bioclimatic conditions do not allow a speedy recovery of vegetation.

A. Marini and M. Talbi (eds.), *Desertification and Risk Analysis Using High and Medium Resolution Satellite Data*,

In Europe the measures to combat desertification include policies and technical actions to by applied on areas affected by current or past desertification processes.

Recommendations aimed at combating desertification need to integrate socio-economic and environmental factors. Actions for making compatible environmental conservation and economical progress should be developed and properly implemented.

Particularly, the new thematic strategy for protecting soils in Europe includes a policy to combat desertification.

Adopted by the European Commission the 22nd of September 2006 it is made up by three components:

The Thematic Strategy for Soil Protection consists of a Communication from the Commission to the other European Institutions, a proposal for a framework Directive (a European law), and an Impact Assessment (Web site EC).

- The *Communication (COM(2006) 231)* sets the frame. It explains why further action is needed to ensure a high level of soil protection, sets the overall objective of the Strategy and explains what kind of measures must be taken. It establishes a ten-year work program for the European Commission.
- The proposal for a framework *Directive (COM(2006) 232)* sets out common principles for protecting soils across the EU. Within this common framework, the EU Member States will be in a position to decide how best to protect soil and how to use it in a sustainable way on their own territory.
- The Impact Assessment *(SEC (2006) 1165* and *SEC(2006) 620)* contains an analysis of the economic, social and environmental impacts of the different options that were considered in the preparatory phase of the strategy and of the measures finally retained by the Commission.

Soil is defined as the top layer of the earth's crust. It is formed by mineral particles, organic matter, water, air and living organisms. It is in fact an extremely complex, variable and living medium. The interface between the earth, the air and the water, soil is a non-renewable resource, which performs many vital functions: food and other biomass production, storage, filtration and transformation of many substances including water, carbon, nitrogen. Soil has a role as a habitat and gene pool, serves as a platform for human activities, landscape and heritage and acts as a provider of raw materials. These functions are worthy of protection because of their socio-economic as well as environmental importance (Web site JRC).

2 Goals

- Preventing further soil degradation and preserving its functions:
  - When soil is used and its functions are exploited, **action has to be taken on soil use and management patterns**.
  - When soil acts as a sink/receptor of the effects of human activities or environmental phenomena, **action has to be taken at source**.
- **Restoring** degraded soils to a level of functionality consistent at least with current and intended use, thus also considering the cost implications of the restoration of soil.

Erosion, loss of organic matter, compaction, salinization, landslides, contamination, sealing... Soil degradation is accelerating, with negative effects on human health, natural ecosystems and climate change, as well as on our economy. At the moment, only nine EU Member States have specific legislation on soil protection (especially on contamination).

The EU Levels of intervention are:

- Soil degradation affects other environmental areas
- Distortion of the functioning of the internal market
- Transboundary impact
- Food safety
- International dimension

Actions and means:

- **Framework** legislation with protection and sustainable use of soil as its principal aim
- Integration of soil protection in the formulation and implementation of national and Community policies
- Research supported by Community and national research programmes
- Public **awareness** of the need to protect soil

For example: Article 1 – Subject-matter and scope

Soil defined as the top layer of the earth's crust situated between the bedrock and the surface, excluding groundwater.

Soil functions to be protected:

1. Biomass production, including in agriculture and forestry
2. Storing, filtering and transforming nutrients, substances and water
3. Biodiversity pool, such as habitats, species and genes
4. Physical and cultural environment for humans and human activities
5. Source of raw materials
6. Acting as carbon pool
7. Archive of geological and archaeological heritage

Different EU policies (for instance on water, waste, chemicals, industrial pollution prevention, nature protection, pesticides, agriculture) are contributing to soil protection. But as these policies have other aims and other scopes of action, they are not sufficient to ensure an adequate level of protection for all soil in Europe.

The strategy is one of seven Thematic Strategies that the Commission has presented. The other strategies cover air pollution, the marine environment, waste prevention and recycling, natural resources, the urban environment and pesticides.

3 Policy options for prevention and remediation in the thematic strategy for soil protection

Soil erosion is a common problem in the European Union. Therefore, although the aim to reduce soil erosion is common, the answers lie in national solutions adapted to each country's needs.

Desertification is land degradation in arid, semi-arid and dry sub-humid areas resulting from various factors, including climatic variations and human activities.

European Mediterranean countries have been identified as sensitive areas to desertification because of the occurrence of particular conditions over large areas:

- Semi-arid climatic conditions affecting large areas, seasonal droughts, very high rainfall variability and high-intensity rainfall
- Poor and highly erodible soils, prone to develop surface crusts
- Uneven relief with steep slopes and very diversified landscapes
- Extensive forest coverage losses due to frequent wildfires
- Crisis conditions in traditional agriculture with associated land abandonment by rural populations and deterioration of soil and water conservation structures
- Overgrazing
- Unsustainable exploitation of water resources leading to serious environmental damage including chemical pollution, salinization and exhaustion of aquifers
- Concentration of economic activity in coastal areas as result of urban growth, industrial activities, tourism and irrigated agriculture

Land degradation has not spatial and temporal confinement. In Northern Europe under northern boreal and sub arctic climate conditions there are examples, which are similar with desertification known from Southern Europe (European Commission, 2004).

Drivers of desertification always include both human and biophysical factors. Land management practices and land use changes leading to overgrazing, deforestation, forest fires, and secondary salinization are among the most recognized causes of land degradation and desertification.

The assessment of sensitivity to desertification and actual desertification itself is an issue addressed by European scientific research projects whose importance increased after the United Nation Convention to Combat Drought and Desertification (UNCCD) committed affected countries to elaborate and approve a National Action Programmes (NAPs).

Desertification is a complex process of simultaneous degradation of soil, water resources and vegetation, which can affect natural, semi-natural and agricultural systems, as well as other human activities.

Collectively, this degradation leads to a loss in resilience, soil quality and in ecosystem integrity and health.

This results in a loss of both ecological and social capital.

Desertion and desertification are related problems in rural areas. Whether causes of land abandonment are natural or socioeconomic is still a subject to debate.

Land abandonment occurs because of external driving forces, such as market changes, or as a consequence of land degradation, which lead the system to cross some irreversible threshold, such as the critical soil depth for plant growth.

Aridity, drought and desertification are distinct, but closely related, concepts. In addition to natural irregularities on water supply associated to climatic conditions, public perception of desertification in Mediterranean areas has been heightened by water resources shortage arising from human induced water problems (Fig. 1).

Successful programmes to address desertification should encompass all the complexity of this problem, including its physical, ecological, sociological and economical components. Moreover, they require reliable instruments of diagnosis and forecast, allowing for the application of the right treatments at the right places.

Combating desertification includes the application of sustainable systems of exploitation of land resources and ranges from the prevention and/or reduction of land degradation to the restoration of degraded/deserted lands.

Prevention and reduction of land degradation should be tackled through linked forecast and integrated land use planning actions. The implications for management of these actions should be refined at the local scale.

Once some thresholds are exceeded, even if desertification-driving forces are reduced, degradation can only be reversed by restoration actions. Since degraded systems are characterized by net losses of resources, restoration in desertification-prone areas is conceived to increase the conservation and capture of such resources. These goals should be achieved by restoring ecosystem functioning.

To avoid, reduce, limit and mitigate land degradation and desertification it is needed to adapt both research and policies to the many variable scales at which socio-natural landscape dynamic operate. More recognition should be given to the importance and responsibilities of all social sectors involved in the study and control of the desertification processes (Annex 1).

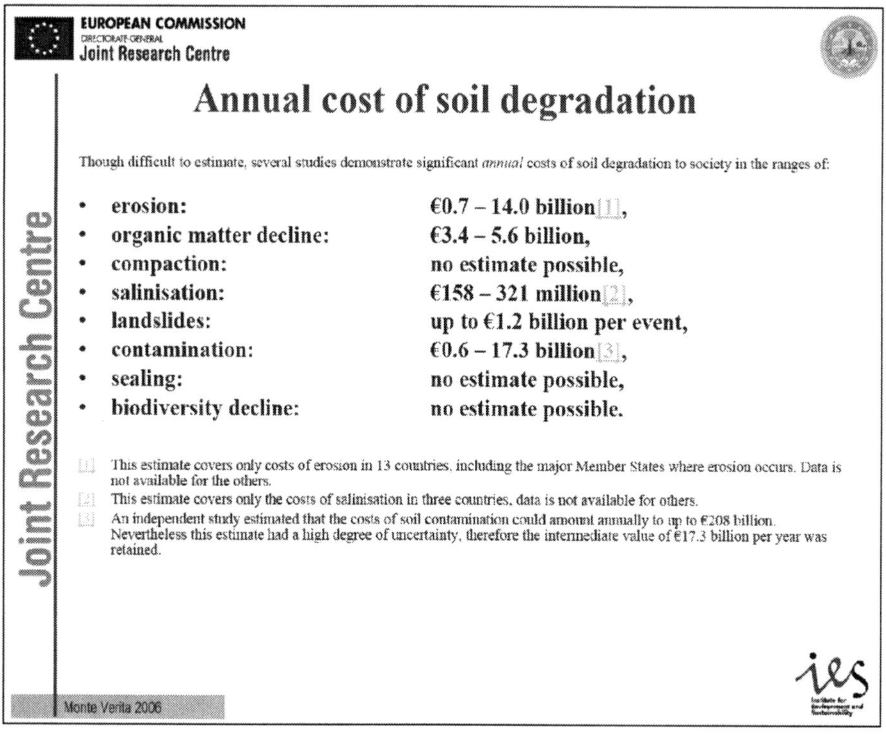

Fig. 1: Annual cost of soil degradation (From Montanarella, JRC 2006)

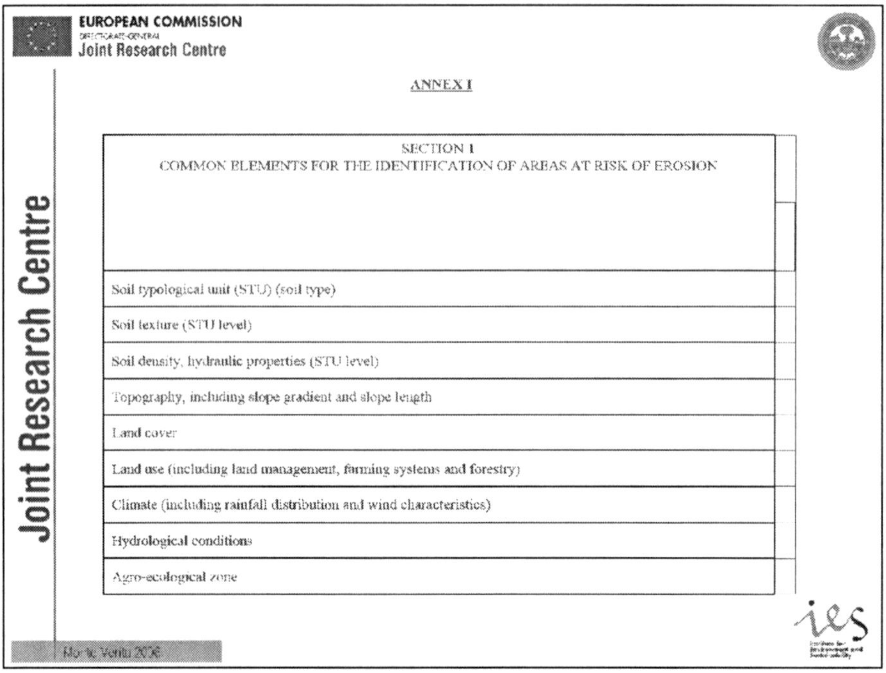

SECTION 1 COMMON ELEMENTS FOR THE IDENTIFICATION OF AREAS AT RISK OF EROSION
Soil typological unit (STU) (soil type)
Soil texture (STU level)
Soil density, hydraulic properties (STU level)
Topography, including slope gradient and slope length
Land cover
Land use (including land management, farming systems and forestry)
Climate (including rainfall distribution and wind characteristics)
Hydrological conditions
Agro-ecological zone

Annex 1: Common elements for identification of areas at risk of erosion (From Montanarella, JRC 2006)

Collaboration and co-operation between European organizations and the countries Mediterranean Neighbors (ENPI programme) should be established in order to promote initiatives concerning joint programmes, financial aspects and technology transfer for developing pilot research projects on soil degradation processes and on mitigation measures.

Bibliography

Web site EC: http://ec.europa.eu/environment/soil/index.htm

Web site JRC: http://eusoils.jrc.it/ESDB_Archive/eusoils_docs/doc.html#OtherReports

L. Montanarella. 2006. JRC Report to soil conference, Brussels.

European Commission. 2004. Reports of technical working groups: thematic soil strategy, vol. II erosion, EUR 21319 EN/2, http://ec.europa.eu/environment/soil/pdf/vol2.pdf.

APPLICATION OF REMOTE SENSING TECHNIQUES TO DESERTIFICATION MAPPING

AEOLIAN MORPHOGENESIS AND STRATEGY OF FIGHT AGAINST DESERTIFICATION IN ALGERIA (HODNA AND ZIBANS BASIN)

L. BOUREBOUNE AND M.T. BENAZZOUZ
University Mentouri Constantine, Algeria
lambourbon@yahoo.fr; mtbenazzour@yahoo.fr

Abstract

The evolution of the relief in the arid areas is under the influence of the wind whose varied actions are expressed by the corrosion exerted by the sand winds, by dune constructions, which can take imposing extensions in the form of ergs.

The action of the wind causes to the engineers problems of fixing of sands because the vegetable cover is here, unimportant to ensure a good stability of the dunes.

The knowledge of the methods of the accumulation of sands requires systematic statements on the ground, which are often lacking in Algeria and in the arid areas of the world.

The fight against mobile sands and the projection of the dunes supposes the precise knowledge of the methods of transport of sands according to the configuration of the ground, of the granulometry of sands, the speed but also direction of the dominant winds.

Various approaches for the fixing of the mobile dunes were often practiced without success; these techniques are defined according to concepts and principles based on the methods of the mobilization of sands by the wind and consist of actions aiming at:

- Forcing the deposit of sands on the upstream wind of the area to be protected
- Increasing the potential of transport of sands by the wind through the zone to protect in order to facilitate a good transit of sands
- Reducing supply of sands to its simple form
- Deflecting mobile sands beyond the surfaces to be protected

These various approaches made it possible to develop and apply the stability techniques of the mobile dunes.

In the light of all these methods, we present some elements of a strategy of fixing the mobile dunes in Algeria, starting from the Hodna basin and Zibans range located at the Eastern ending of the Saharian Atlas.

In the Hodna basin and the area of Zibans we identified fields of sharp dunes whose current progression threatens stranding cities like Bou Saada, the town of Biskra and the oases in Zibans. The preliminary treatment of these mobile dunes showed a failure because of specific interventions, very localized; the fixing of

A. Marini and M. Talbi (eds.), *Desertification and Risk Analysis Using High and Medium Resolution Satellite Data*,

the mobile dunes must pass through the study of wind dynamics within its global framework in order to be able to determine the extension of the sphere of activity of the wind; zones of supply sands, aeolian corridors or of transit of sands and accumulation areas.

On the basis of control and identification of these zones of actions of the wind, one can advance suggestions for a solution to the effective treatment of the mobile dunes and their fixing in these areas in the framework of the strategy of the fight against desertification in Algeria.

Keywords: Desert degradation, sahara, wind erosion, mobile dunes, drifting sand, aeolian corridor, green dam

1 Introduction

The sand movements appear in various environments, however, they take a great importance in the arid zones because they are at the origin of serious problems of stranding which affect cities, infrastructures of transport, oases and cultivated areas.

In the arid zones, the sites of the cities, oases, can constitute obstacles and disturbances of the mechanisms of preexistent Aeolian morphogenesis.

In order to avoid and to reduce the harmful effects of mobile sands, it is necessary to know the nature of mobilization of sands, to identify the types of actions which tend to accelerate the process and the practical measurements which can reduce its actions.

2 The nature of sands and their mobility by the wind

2.1 Granularity of transported sands

The mobility of sands depends on five factors: swiftness, the turbulence of the wind and characteristics of the surface of the ground (Fryberger, 1979).

The interaction between the erosive capacities of the wind and the erodibility of the surface of the ground determines if sand is mobilized or not. The role of the vegetation is not negligible; it is a disturbing factor, which reduces the speed of the surface wind (Buckley, 1987; Ash and Wasson, 1983).

The granulometry of sands varies from 0.08 to 2.00 mm and its mobility is carried out at ground level by saltation.

A first limit is the minimum threshold of dimension from which it is considered that the particles are aeolian sands and not dust; threshold proposed by J. Dubief indicates value of 1/15 mm (65 μm) while R.A. Bagnold does not admit that sands higher than 1/5 mm are 200 μm with 1/4 mm (250 μm), he then considers, like truly sandy that the particles transported by saltation or bearing.

However the data on sands dune express a granularity whose mode is located between 160 and 320 μm, so, we retain the value of R.A. Bagnold by decreasing them slightly.

2.2 Methods of aeolian dynamic

The mobilization of sands fits perfectly in the mechanisms of aeolian dynamics through its three fundamental processes that are the deflation, the transport and the accumulation.

When the speed of the wind is constant, then it is established a balance between the relative flow of the air and sand. However, any disturbance of the speed of the wind breaks this balance in the case of either by starting ablation acceleration, or in favor of the deposit in the case of a reduction in the force of the wind (Wasson and Nanninga Token Entry, 1986).

2.2.1 Deflation

Deflation consists of an individual removal of particles by the wind; deflation affects especially stripped and dry surfaces.

While *corrosion* results in a mechanical action exerted by winds violent ones strongly charged with abrasives, like quartzy sand.

2.2.2 Aeolian transport

The wind is a very effective freight agent of sands. Even if its competence is limited to the low-size elements because of the low density of the air, it ensures the transport of considerable tonnages of materials.

There are three means of transport of sand in the air proposed and described by R.A. Bagnold (1941) and which are unanimously allowed:

- Transport by suspension (the particle floats in the air)
- Transport by saltation (the particle carries out jumps in the air)
- Transport by rolling (the particle rolls on the ground)

For the sandy particles composing the dunes, the saltation is the mean transport mode; the sand grains draw a parabolic trajectory of form before falling down on the ground.

The saltation or jump by successive jumps represents the most important mode of displacement of sands, one estimate that it ensures the 3/4 of wind transport.

Displacement by successive jumps defines trajectories being able to reach 1.50 m height and a 2 m length and engages especially medium sands up to 0.5 mm, that is to say 500 µm.

2.2.3 The deposit or the accumulation of sands

The deposit of sand is carried out when the wind ceases or falls at a lower speed limit from the one necessary to transport.

Accumulation translates a perfect homometry of the deposit, which shows well sorting ensured by the wind. One can recognize a typical intersected structure of deposit of dunes.

The movement of sand is carried out when the wind reached a higher speed or equalizes with 6 m/s, concept of effective wind.

I. Williams shows that there is deposit when the current wind is at the same time saturated with sand and slow moving or convergent in the horizontal plane.

If saturation and the deceleration are carried out, sand deposits in any site.

As for the problem of the deposit, there remains related to that of the progression of the dunes; in fact, an active dune progresses in the absence or in the presence of vegetation if this one does not cover it completely.

If the source of sand and the wind able to drive it exist, the only condition necessary to the deposit is the cause of accumulation. There remains clear that this condition can be:

⇒ A biological factor, the vegetable cover or
⇒ A topographic factor: a sandy surface, or a rough surface or a wet surface
⇒ Or a mechanical factor, which can present several forms:

- A loss of competence of the wind: it is the phenomenon of alluvial deposit for J.P. Gallois who defines it as:

"the speed of the wind decreases, so, the saltation occurs by a smaller quantity of sand and the number of grains which stop is higher than the number of those which continue"; it can be also a combination of a topographic element and of a mechanical data, an important obstacle which provoke the phenomenon "*of encroachment*" for J.P. Gallois, when it occurs a brutal topographic lowering in hollow or relief, reptation mechanism is stopped, while the saltation continues and causes an accumulation.

2.3 Potentialities of wind erosion in Hodna basin and the Zibans range

To the mechanisms of wind erosion, developed above correspond of the specific surfaces:

The surfaces with dominant of supply of material
The surfaces with dominant of transport or material in transit
The surfaces with dominant of reception, alluvial deposit or accumulation of material

It should be specified that this last sand category might function later on in starting surfaces where source areas of sands previously deposited.

Any sand departure can be carried out only if the wind tears off initially the particles of the ground before moving them at variable distances.

In other words, the sand starting surfaces represent in fact weakened zones which lost their protection or cohesion in the form of vegetable cover more or less dense or in the form of soils, following various degradations such as overgrazing, land under cultivation…

Enormous sand stocks covered the valleys, plains and slopes through Hodna basin and the Ziban range, and presented then, the existence of a high potential of mobilization of sands just that the latter do not have any more cohesion (Fig. 1).

Our investigations on the field, enabled us to lead to the discovery of three types of possible origin of sands (Benazzouz, 2000) which are:

⇒ Wadis where sands are systematically recycled on bank oriented to the wind.

⇒ Holocene erg, with more or less vegetation but especially showing an advanced state of degradation.

⇒ The trans-atlas aeolian system is composed by two systems: Zahrez-Hodna and Hodna-Zibans, it is the true vector of displacement of sands from the North-West towards South-East all along its corridor on the ground, materialized by sandy dunes and wind veils more or less thick.

Fig. 1: The trans-atlas aeolian system in Zahrez-Hodna section, djebel Kenfoud pass invaded by mobile sands coming from West, in Zahrez Chergui basin. Forms of fight against the mobile dunes

3 Forms of fight against the mobile dunes

3.1 Principles and design of fight against the stranding

Mobile sands can be at the origin of serious problems as soon as there is reduction in the force of the wind or a change of the texture of the surface of the ground. Then phenomena of stranding and accumulation of sands occur, which seriously threaten the infrastructures of transport and other installations in these areas.

Any fight against the misdeeds of the stranding is based on the principles and following designs:

3.1.1 To facilitate and cause the accumulation of sands upstream wind

The objective consists in reducing the transport of sand by the wind while acting on the competence of this one.

- To make pits from 3–4 m length (distance covered by the grains during the saltation) in upstream wind of the zone to be protected.

- Barriers and hedges: types of obstacles against the wind creating a reducing surface of the transport capacity of the wind ahead and behind of the obstacle.

3.1.2 To reduce the sand supply

When the rates of accumulation of sands are high behind the edges, it is necessary to seek to reduce the additional sand contributions while acting on the source areas of supply sands by techniques of treatment of the surface of the ground.

3.1.3 The deviation of mobile sands

In order to avoid the stranding of infrastructures, a recommended method from Aramco, consists in deviating the sandy currents which will circumvent the installations to be protected instead of attacking them on the front.

3.2 The techniques of fixing mobile dunes

Among these techniques, we will retain the following as the most current:

3.2.1 The rehandling of the dunes

In order to protect installations like roads, railways, pipelines, one can act on the destruction and the rehandling of the dunes.

The method consists of a mechanical excavation of the dune and the massive transport of sands towards a new place.

3.2.2 The dissipation of sandy dunes and destruction of the dune

It is a question of modifying the shape of the dune by replofilage, to treat the surface of the dune by sprinkling it of chemicals or hydrocarbons products.

3.2.3 The fixing and stabilization of the mobile dunes

The present new method consists to level the dunes and to sprinkle topography, thus leveled by using fixing products derived from oil (National Iranian oil Company), this chemical spreading strongly fixes sand, makes it possible to retain moisture.

3.3 The delimitation of the functional wind corridors in the Hodna basin and the Zibans range

From the point of view of the geomorphologist, the fight against the stranding and the fixing of the mobile dunes pass by the perfect knowledge of particularly active wind morphogenesis in the margins of the desert.

It is acted in fact, to carry out the delimitation of a certain number of particularly active wind corridors by where transit forwards the main part of the sandy masses. These sites were the subject of a study detailed through Hodna and Zibans (Benazzouz, 2000), of which we consider appropriate to report here some complementary data.

3.3.1 The wadi Maiter gap and the Bou Saada plain

The wadi Maiter gap is similar, with pseudo moved back, to the cliff of the djebel Gorehor, opened towards the East in direction of the plain of Bou Saada.

This gap lengthens from West to East on 10 km and develops a width from 2 to 3 km, it is closed in the west by the djebel Kenfoud (Figs. 2 and 3).

The originality of this valley lies in the fact that it is completely invaded by the sands mainly coming by the djebel Kenfoud pass in the West.

The result is the Southern side of the djebel Kenfoud that disappears under a true sand fall (Fig. 4) or "cascade of sands" flowing in the East of the Temsa wadi, because this one is lost downstream under important sand masses in which the Maiter wadi registers its bed.

The junction of the two wind gaps or corridors consisted on the valleys of the wadi Temsa in the West and the Maiter wadi in the East, takes place on the southern side of the djebel Kenfoud which ensures, here, an important and effective transit of sands in the wind corridors.

The gap of wadi Maiter is a particularly active wind corridor where the trains of barkhanes regularly follow one another in their progression towards the East, then threatening dangerously the town of Bou Saada of stranding.

The fatal error reveals a scientific ignorance of the phenomenon of the stranding. Indeed, the gap of wadi Maiter functions as a wind system, admitting a surface of supply sand from the West on the side of the djebel Kenfoud, and a zone of transit at the bottom of the valley of the Maiter wadi.

It is because the configuration of this wind corridor takes the form of one gap opening towards the East in the plain of Bou Saada and until the djebel Meharga that is the zone of transit of sands; it is overflowing largely and the barkhanes dunes progress further towards the East.

Unfortunately, the town of Bou Saada is located exactly in the zone of transit of sands, and any treatment against the stranding here is bound to failure.

On the other hand, taking actions in order to slow down and fix sands which cover the djebel Kenfoud could probably deprive the zone of transit downstream from considerable sand masses and would largely affect the flood of sands in transit.

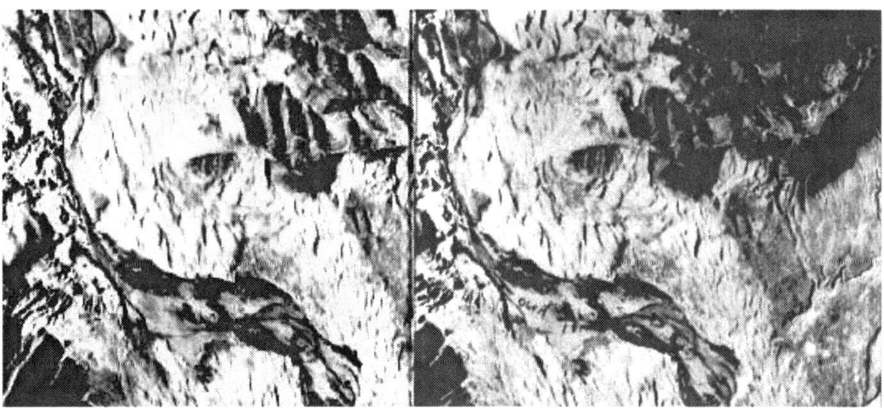

Figs. 2 and 3: Sand transit on the djebel Kenfoud and invasion of the valley of the Temsa wadi

Fig. 4: Sand falls or cascade of sands recovered djebel Kenfoud

Only at this stage of work one could consider the stabilization of the barkhanes at the location of Bou Saada town.

3.3.2 The corridor of M'Doukal

The configuration of the relief translates a broad opening of the Atlas Mountain in the area of M'Doukal town, where the reliefs are draped with aeolian sands on their slopes exposed towards North and the Northwest (Benazzouz, 1994).

This situation confirms the impact of the trans-atlas aeolian flow which operates between the Hodna basin and the Ziban range, the aeolian system developed and analyzed being higher.

In fact, the wind corridor of M'Doukal breaks up, in two segments:

In the North-West the very active corridor of Meksem Fozna fits between the djebel Fozna and the djebel Sahbana into the South and is characterized by a series of small sharp ergs, skirting the northern slope of the djebel Sahbana and often with a thick sandy veils covering the plain with fields of barkhanes.

In the North, an alignment of small dunes, constitutes an important sand stock within this wind corridor.

3.3.3 The wind corridor of the El Outaya plain

In the plain of El Outaya, one meets the vastest fields of dunes of all the area, centered in the alluvial fan of the Selsou wadi.

One can follow the field dunes over a length 20–25 km from North to the South, whereas their width can reach 10 km.

Several generations of ergs accumulate in this vast sandy mass from Holocene ergs with parabolic dunes to the current erg with mobiles dunes.

It is clear that the existence of the wind corridor in the North-western – South-east axes is largely facilitates the recycling of the sands transported by the various generators of the alluvial fan of the Selsou wadi (Fig. 5), in fields of dunes directed towards South-east.

Any development uncontrolled in the place of the plain is bound to a fatal stranding and determines the transformation of this perimeter into a field of dunes, as already occurred to Bled Magraoua.

Fig. 5: Wadi Selsou at the top of the alluvial fan: sand arrival from the risings of the wadi and recycling of sands with formation of dunes on the slopes

3.3.4 The Saharan Piedmont of Zibans

The absence of relief through Saharan Piedmont justifies the full development of the wind corridor which affects the whole of Zab Rharbi situated at the West of Biskra town.

The wind action within this corridor results in the stranding of oasis, as well as in a current progression to the South at front of the barkhanes dunes (Benazzouz, 2000).

Indeed, the palm plantations of Zibans are considered as obstacles located inside the wind corridors that we thus defined on axes of sand transport and transit.

The oasis of Zibans are thus a privileged place of deposits of sand and the many abandoned oasis that are due to stranding can confirm it.

This corridor extends in fact, the impact of aeolian trans-atlas flow towards the South and of which the effects are still perceptible until edge of the chott Merouane.

The origin of sands in this corridor is double:

- Local supply, by the sandy wadis and veils
- An important sand contribution starting from the plain of El Outaya, sands which one finds in the form of intramontane plugs on the heights of the djebel Bou Rhézal (Fig. 6).

Thus, aeolian morphogenesis functions inside a distinct geomorphological unit and trans-atlas wind flow become effective on the ground when it is exerted and took directly on such units.

Fig. 6: Sand dune on top of mountains as intra-mountain blow sand trappings on the northern side of djebel Bou Rhezal at the West of Biskra

In addition, the wind system that we defined checks on Saharan Piedmont where we find a zone of interconnection intra cells or intra wind corridors which makes it possible mobile sands to be able to forward and progress towards the South.

4 The failure of the "Green DAM" in Algeria

4.1 The direction of the dominant winds

At the time of its design, the fundamental error of the green dam is not to have taken account of the direction of the dominant winds through the Northern margins of the desert.

We had announced (Benazzouz, 1993, 1996) that the prevalent direction of the winds on the High Algerian Plains is always influenced by the strong activity of the flow from the Northwest, confirmed throughout the Saharan Atlas, of West in East at the borders of the Moroccan frontier, Ain Sefra (P/year: 114 mm), the resultant of the winds is Northwest-West.

On southern Piedmont of the Atlas, the prevalent direction of the winds with El Abiod Sidi Sheik is from the Northern sector to the North-West.

In the high plains of the Center, in Djelfa (P/year: 335 mm) the winds blow from the West and the Northwest.

In the East, in the Hodna basin, the winds blow in the West and the Northwest at Bou Saada town (P/year: 270 mm).

In addition, the analysis of the diurnal variations indicates that the winds of the West and from Northwest act during the middle of the day and at the beginning of the evening, over a desiccated ground by removing the fine particles of the ground. The morphogenic actions of these winds are very intense in autumn on a ground practically desiccated and made pulverulent.

Consequently, the effective action of the winds being in the Northern sector and in the Northwest: they act directly on sandy grounds destructed and deprived of covering vegetation and this, on a large scale occurs through the High Plains.

Thus, the winds effective and likely to mobilize sands blow of the Northwest in all Eastern Algeria and Tunisia, whereas the principal goal of the Algerian green dam consists in stopping the invasion of sands coming from the Sahara i.e. exactly contrary to the reality of the direction of the effective winds.

4.2 The reverse direction of the Algerian green dam compared with the zones of action of the wind

The shift towards the zones of actions of the most obvious wind is remarkable in the basin of Hodna where the green dam is based on North on the surrounding relief, avoids and surrounded completely the basin of Hodna by the North, leaving a large gap which opens towards South-east and the South, then facilitating largely the easy circulation of the dominant winds from the North-West.

Thus the green dam does not constitute an obstacle facing the sands in the Hodna basin, and it does not offer effective protection against the threat of the desertification.

This lack, as remarked, of the green dam facing the zones of action of the wind is well illustrated in the Zibans range which materializes the large gap of the green dam where we showed the effectiveness of wind morphogenesis, with a strong aggressiveness of mobile sands which not only cross the plains and bottoms valleys, but especially cross the reliefs to further continue their transit towards the South, on Saharan Piedmont and even beyond that.

Thus all these forms of maladjustment raise and betray a vision more technocratic than scientific of green dam in Algeria, coherent with the plantation not by the local populations, but by the Army with its large manpower of young soldiers of the National Service, free altogether, coherent labor with a vision instrumentalist of the fight against the desertification.

In spite of these failures, the green dam is not therefore useless, since some cases of successes can be mentioned:

- Some place of reconstitution of the forests relics on the tops of slopes in the areas of Tébessa, Djelfa
- Fixings of mobile dunes in some test areas on the dune field of Zahrez, the area of Bou Saada, and Naama towns

These some cases of success of the Algerian green dam take part in the maintenance of the grounds, decreasing the erosion of the grounds by the stop of the eradication of the ligneous plants in those zones often strongly degraded.

In a word, they are minor forms of degraded ground rehabilitation (Le Houerou Nr, 1992), actions mentioned in the objectives of the green dam, but very localized and specific even often partial, unimportant achievements compared to the gigantism of the objectives of the Algerian green dam.

5 Conclusions

The fight against desertification in Algeria is also a fight against the stranding and the study of aeolian morphogenesis, and it is necessary to better know the phenomenon in order to be able to stop it (Benazzouz, 1997).

The origin of these forms of stranding is to be sought beyond the affected surfaces, here, the oases of Zibans and the town of Bou Saada as local examples, upstream wind of the conveying wind currents. For that, the regional scale allows a better approach of the dynamics of the dune buildings and of invasion, often revealing the directions of contribution of sands and making it possible to consider suitable solutions.

But the local scale should not be neglected for as much; it enabled us to highlight the existence of the mechanisms of relay of wind cells in a continuum which makes it possible to feed out of sand the whole of Saharan Piedmont.

As a conclusion, any proposal for a fight against the mobile dunes should necessarily consider three phases:

⇒ *1st phase:* exact delimitation of the wind corridor and its extension
⇒ *2nd phase:* priority intervention always with the upstream wind i.e. upstream of the wind corridor, followed treatment of the zone of junction
⇒ *3rd phase:* treatment within the wind corridor, with direct intervention on the dunes by various specified techniques (Cf. above)

The area of Hodna and Zibans is particularly interesting for the study of desertification: the environment is affected by rapid paleoclimatic variations and by a massive intervention of the human activities.

Admittedly, if the assessment reveals the existence of forms of degradation of the environment with stranding more or less generalized of many oases on Saharan Piedmont, one cannot allot them yet to anthropogenic actions or to causes of climatic origin within the framework of aridification.

However the generalization of the current agricultural development, the over-grazing and a strong human pressure in Zibans are at the origin of an rapid progression of the stranding which threatens the oases of Saharan Piedmont.

As for the green dam, intended to fight against the desertification, its effectiveness is questioned locally in Hodna and Zibans, where wind flow transatlasic generating effective winds of the North-West, acts largely as direction of the Sahara, whereas the green dam is located completely at North, circumventing a vast active zone of migrations of sands and progression of forms of stranding towards the South.

The implication of the local populations, strong with their traditional knowledge, of their adaptation to the arid environment, knowing perfectly from which the wind comes, and knowing where to carry out their herd, constitutes the major axis which will guide any policy of installation and fight against the desertification in these particularly fragile semi-arid zones in Algeria.

Bibliography

Ash J.E. and Wasson R.J. 1983. *Vegetation and sand mobility in the deserted Australian dunefield*, in Zeitschrift für Geomorphology, Sup., n.**45**, 7–25.

Bagnold R.A. 1941. *Deserted the physics off blown sands and dunes*. Methuen, London, 265 p.

Benazzouz M.T. 1993. *L'endoréisme dans les hautes Plaines algériennes: Origine et impact dans les mécanismes de la désertification*, In colloque en l'honneur du Professeur R. Coque. Paris 11 et 12 février 1993: Les milieux arides et semi-arides: héritages et dynamiques actuelles. Publication du Centre de Biogéographie et Ecologie, ENS Fontenay Saint-Cloud.

Benazzouz M.T. 1994. *Etude des interactions relief-migrations éoliennes de sables dans la région de M'Doukal (Algérie)*, rev. Méditerranée n.**3–4**, pp 51–58.

Benazzouz M.T. 1996. *The causes of desertification in the northern Algerian Sahara*, in Geomorphic hazards, edited by Slymaker O., Wiley, 183–195.

Benazzouz M.T. 1997. *Morphogénèse éolienne et désertification dans les monts des Zibans, Algérie*. Fourth International Conference on Geomorphology, Bologna, Italy, 28 August–3 September 1997.

Benazzouz M.T. 2000. *Morphogénèse éolienne holocène et actuelle dans l'Atlas saharien oriental: conséquences sur la désertification*. Thèse Doctorat d'Etat, université de Constantine, 350 p.

Buckley R. 1987. *The effect off sparse vegetation one the transport off dune by wind*, in Nature, n.**325**, 426–428.

Fryberger S.G. 1979. *Dune forms and wind mode. In A study off total sand sea*, edited by Mc Kee U.S., Geological Prof Paper, **1052**, 137–169.

The Houerou H. Nr 1992. *Biogeographic Research on the steppes of the North of Africa.* Thesis doctorate of State, University Paul Valéry – Montpellier, 1992.

Wasson R.J. and Nanninga Token Entry. 1986. *Estimating wind transport off sand one vegetated surfaces*, in Earth Surfaces Processes and Landforms, n.**11**, 505–514.

Welshman J.P. 1965. *The fight counters the stranding, review of inform.* Saharan organization, n.**17**, Algiers, 13 p.

Wilson I.G. 1971. *Sand waves. Universal discontinuities in bedforms produced by the wind,* in New Science, n.**23**, 634–637.

ESTIMATING SOIL MOISTURE USING OPTICAL AND RADAR SATELLITE REMOTE SENSING DATA

STEFANO NATALI[1], LORETO PELLEGRINI[1], GIANLUIGI ROSSI[3], LUDOVICA GIORDANO[2], MASSIMO IANNETTA[2], GABRIELE SCHINO[2], ALBERTO MARINI[3], AND GASMI NABIL[3]

[1]*MEEO S.n.c. San Giovanni di Ostellato – Ferrara – Italy, natali@meeo.it*
[2]*ENEA CR Casaccia, Italy, miannetta@casaccia.enea.it*
[3]*Università degli Studi di Cagliari, Italy, marini@unica.it*

Abstract

The purpose of this work is to investigate the possible use of MODIS and ENVISAT-ASAR Data to extract soil moisture (SM) field and validate the developed systems by means of *in situ* measurements.

The main idea of this study concerns the advantages that could arise from a continuous and almost real time SM monitoring, permitted by wide-swath passive optical sensors that allow evaluating the dynamic evolution of different phenomena on a wide range of application fields: from hydrological monitoring and management to agricultural applications and so on. Observations provided by MODIS have no direct physical relation with soil water content. On the other hand, the relationship between Land Surface temperature (LST), NDVI and *in situ* measurements have shown many potentialities to indirectly retrieve SM information.

The use of SAR data for validation purposes is envisaged but not yet applied.

Calibration and validation phases are performed over a test site where *in situ* measurements are available.

Keywords: Soil moisture, Thermal Inertia, optical, radar sensor, SoA-TI, MODIS, ASAR

1 Introduction

The estimation and modeling of large-scale ecological processes requires the availability of spatially explicit data on the distribution of key environmental variables. Among the most important of these environmental variables are vegetation cover and soil water content. The latter in particular, plays an important role in ecological processes such as climate change, agriculture, evaporation, transpiration, flood forecasting, surface run-off and ground water replenishment (e.g., Munro et al., 1998).

Obtaining spatially explicit data on vegetation cover and soil water content (i.e., mapping them) on the basis of ground sampling is, however, exceedingly expensive and time consuming, given their wide spatial variability. Mapping has thus relied extensively on satellite remote sensing. However, while a variety of methods have

A. Marini and M. Talbi (eds.), *Desertification and Risk Analysis Using High and Medium Resolution Satellite Data*,

been implemented and tested for estimating and mapping vegetation cover (Tucker, 1979; Verstraete and Pinty, 1996; Turner et al., 2005), much less effort has been directed at estimating and mapping soil moisture on the basis of remotely sensed data.

Soil moisture (SM) is defined as the mass or volume of water by unit mass or volume of dry soil. Given the large difference in dielectric constant between dry soil and water, the dielectric constant of whole soil is highly dependent on soil moisture. Therefore, measurements of bulk dielectric properties, such as those done in the microwave region, are directly related to soil moisture. Given these physical relations, it is not surprising that many efforts in estimating soil moisture have relied on active radar sensors (Dubois et al., 1995; Boisvert et al., 1996; Glenn and Carr, 2004; Alvarez-Mozos et al., 2006), and active microwave data have indeed been demonstrated to depend on several natural surface parameters such as dielectric constant and surface roughness.

More recently Portmann (2000) proposed a system to extract soil moisture information based on Soil-Vegetation-Surface-Transport scheme (SVAT) and Land Surface Temperature (LST) data. This method exploits the indirect signal of soil moisture that can be derived from its influence on the partitioning of net radiation into latent, sensible and soil heat fluxes. The variation of the LST subject to solar irradiation is the key element used in the algorithm. This is because the latter two fluxes are largely dependent on the LST and especially on the steepness of the LST rise at sunrise (8:00–10:00 local solar time), while the shape of the curve has been shown to be quite indicative of soil moisture (Wan et al., 1999).

This study builds on an understanding of the advantages that could arise from the continuous and almost real time SM monitoring permitted by geostationary and wide swath passive optical sensors like MODIS. In this paper, we thus present preliminary results on the comparison of active microwave and passive optical/ infrared data in estimating soil moisture in arid regions of Northern Africa. MODIS-derived SM was used for cross-validation of results derived from the radar system.

2 Data & methods

Satellite data available for analysis included two ASAR images acquired in geo-coded AP (HH/VV) mode and four MODIS images (two diurnal and two nocturnal images) covering part of Northern Africa.

In particular, to retrieve soil moisture information from MODIS data the SoA-TI model (Cai et al., 2005) was selected and applied. The model, first proposed by Ma and Xue (1990), is a bi-temporal three-step system that allows estimating SM content from Thermal Inertia (TI) maps over non-vegetated areas. It requires, as input, two multispectral (daytime and nighttime) MODIS scenes, and computes Albedo and Land Surface Temperature (LST) to generate TI maps (Xue and Crackell, 1995).

$$TI = \frac{(1-A)S_O C_t}{\Delta T\sqrt{\omega}}\left[\frac{A_1\left[\cos(\omega t_2-\delta_1)-\cos(\omega t_1-\delta_1)\right]}{\sqrt{1+\frac{1}{b}+\frac{1}{2b^2}}}+\frac{A_2\left[\cos(\omega t_2-\delta_2)-\cos(\omega t_1-\delta_2)\right]}{\sqrt{2+\frac{\sqrt{2}}{b}+\frac{1}{2b^2}}}\right] \quad (1)$$

The TI field is calculated using (1):

Where:

TI is the Thermal Inertia field
A is the Albedo
S_o represents the solar constant
C_t is the atmospheric transmittance in the visible spectrum
ΔT is the diurnal temperature difference

- Is the Earth's angular frequency

A1, A2 are the Fourier series coefficients

- Is the solar declination
- Is the latitude

For the specific case of the MODIS sensor the Albedo is calculated as proposed by Liang (2000), see Equation (2):

$$\alpha_{Short} = 0{,}160\alpha_1 + 0{,}291\alpha_2 + 0{,}243\alpha_3 + 0{,}116\alpha_4 + 0{,}112\alpha_5 + 0{,}081\alpha_7 - 0{,}0015 \quad (2)$$

Where each α_i represents the reflectance of the MODIS band i.

The diurnal temperature difference represents the maximum day-night temperature difference, and is calculated following the approach suggested by Xue et al. (1995): assuming that the maximum temperature is expected for 14:00 local time and the minimum temperature is expected at 02:00 local time, and assuming that the diurnal temperature variation has a cosine distribution; (3) allows extracting the maximum and minimum daily values based on the daytime and night time satellite measurements; (4) represents the diurnal temperature difference.

$$T_{(14,02)} = T_{31} + A\cos(\omega t_1) \quad (3)$$

$$\Delta T = T_{max} - T_{min} \quad (4)$$

Where:

$$T_{max} = T_{14} + \frac{(T_{14} - T_2)\cos\left[(\omega t_{max}) - \cos(\omega t_1)\right]}{\cos(\omega t_1) - \cos(\omega t_2)}$$

$$T_{min} = T_2 + \frac{(T_{14} - T_2)\cos\left[(\omega t_{min}) - \cos(\omega t_2)\right]}{\cos(\omega t_1) - \cos(\omega t_2)}$$

SM fields are then retrieved from TI maps making use of the Ma and Xue relationship – see Equation (5).

$$TI = \left[2{,}1ds^{(1{,}2-0{,}02\frac{ds}{d}SM)} \cdot e^{(-0{,}007(\frac{SMds}{d}-20)^2)} + ds^{(0{,}8+0{,}02\frac{ds}{d}SM)}\right]^{\frac{1}{2}} \cdot \frac{\frac{0{,}2SM}{d}ds^2}{0.001}\sqrt{100} \quad (5)$$

Where *SM* is the SM field, *d* and d_s represent the mean soil density; Look-Up Tables were created to associate SM to different TI values.

The version here implemented is an improved version of the original SoA-TI method; daytime and night time LST fields were directly extracted from the MODIS L2 LST product, while cloud, water bodies and vegetation land cover were masked by means of the land cover map generated applying SOIL MAPPER® (MEEO, 2007) to the day-time MODIS scene. Figure 1 shows a flow chart of the implemented MODIS-based SM retrieval method.

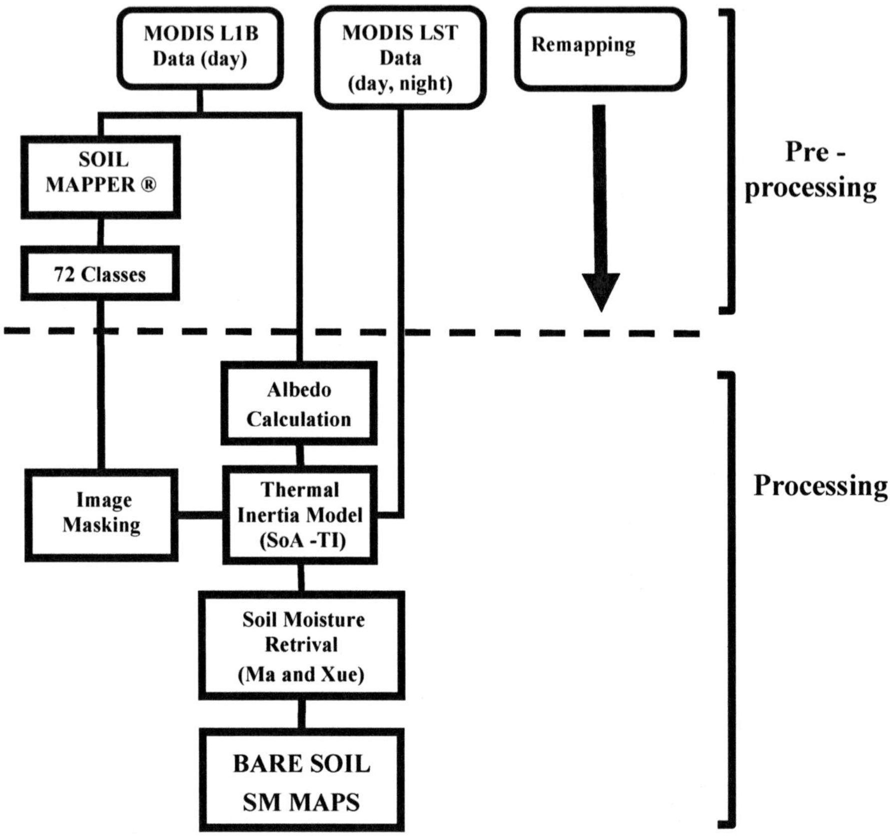

Fig. 1: Schematic view of the MODIS-based soil moisture retrieval method; the pre-processing phases applied to both daytime and night time data are shown

To obtain SM information from ASAR data different empirical and semi-empirical methods were evaluated (Cravey et al., 2008; Bignami et al., 2004; Paloscia et al., 2005; Baghdadi and Zribi, 2006); eventually, it was decided to apply a semi-empirical approach that aims at retrieving a direct relationship between ASAR multi-polarization backscattering coefficient values (σ^0_{HH}, σ^0_{VV}, σ^0_{HV},) and SM values directly measured (in-situ surveys). ASAR data were processed using the BEST software to obtain backscattering maps. MODIS SM-retrieved fields and ASAR backscattering maps were compared remapping MODIS data on a UTM Zone N32 projection and downscaling ASAR high-resolution data to the MODIS 1 km resolution.

Finally six *in situ* soil moisture measurements by gravimetric samples were collected in Tunisia on 15th January 2007, corresponding to the first ASAR data acquisition date (see Table 1).

Table 1: Backscattering coefficients corresponding to ground sampling points

Sample ID	σ_{HH}	σ_{VV}	**Soil moisture (%)**
1A	0.08	0.05	12
1B	0.08	0.05	12.7
1C	0.08	0.05	13.2
2A	0.07	0.25	12.3
2B	0.07	0.25	13.1
2C	0.07	0.25	11.7
3A	0.05	0.12	11.1
3B	0.05	0.12	12
3C	0.05	0.12	11.1

3 Results

3.1 MODIS data

Figure 2 shows the daytime MODIS Band 1 in reflectance units for the first event (15th January 2007) on the target area (northern Africa) (a) and masked for water bodies, vegetation and clouds for both daytime and nighttime.

The retrieved SM Map together with the frequency distribution of soil moisture over the scene for the first event is shown in Fig. 3. Similarly to Figs. 2 and 3, Figs. 4 and 5 show the Band 1 masking and the Soil Moisture distribution for the second sampling event (2nd May 2007) respectively. Note that increased cloud cover reduced the area available for analysis.

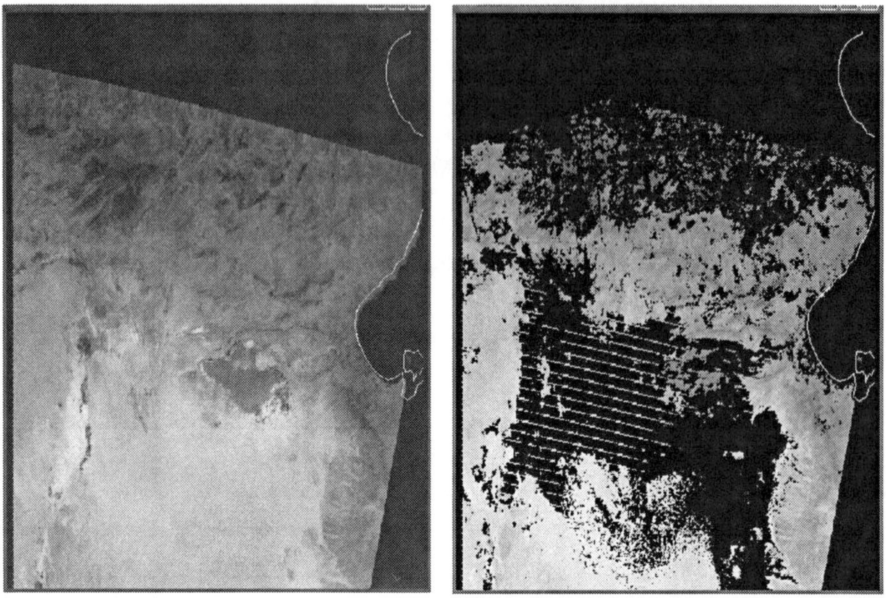

a) b)

Fig. 2: Map of Band 1 reflectance values over the target area for the first event (15th January 2007) with no masking (a) and masked for water bodies, vegetation and daytime-nighttime cloud mask (b)

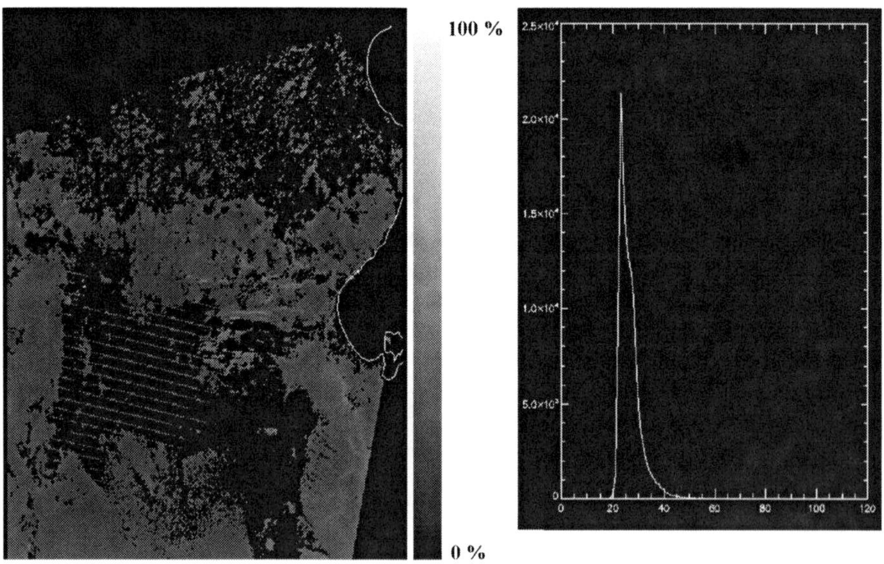

a) b)

Fig. 3: Maps of soil moisture (a) and frequency distribution over the scene relative to the first sampling event (15th January 2007) as derived from MODIS data

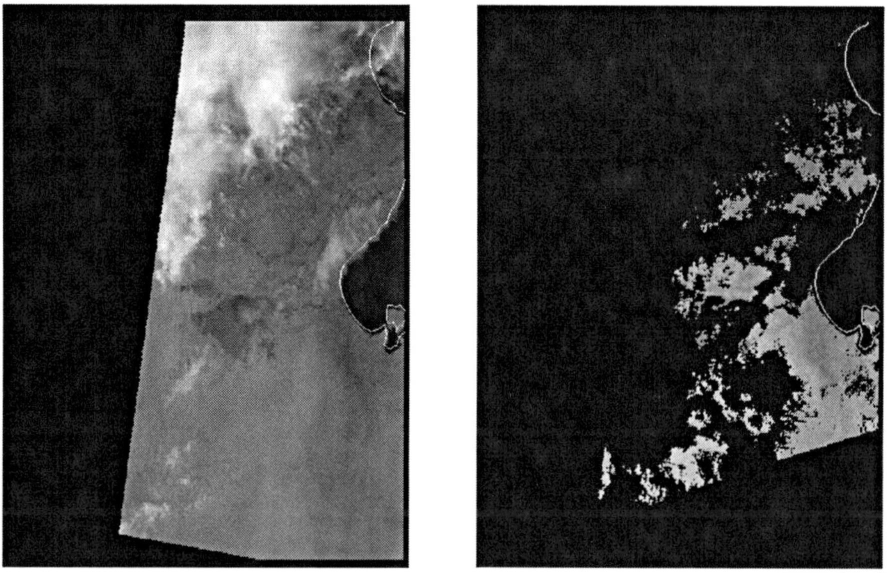

a) b)

Fig. 4: Map of Band 1 reflectance values over the target area for the second sampling event (2nd May 2007) with no masking (a) and masked for water bodies, vegetation and daytime-night time cloud mask (b)

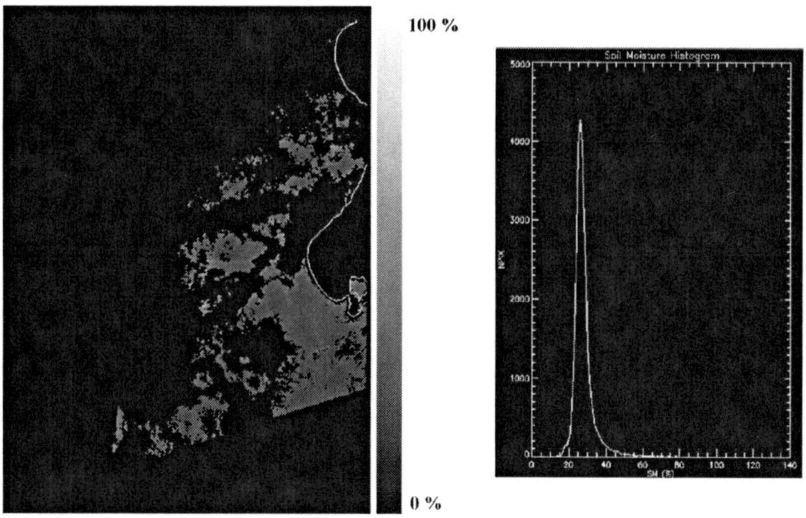

a) b)

Fig. 5: Maps of soil moisture (a) and frequency distribution over the scene relative to the second sampling event (2nd May 2007) as derived from MODIS data

Figures 6 and 7 compare directly the soil moisture estimated during the two sampling events.

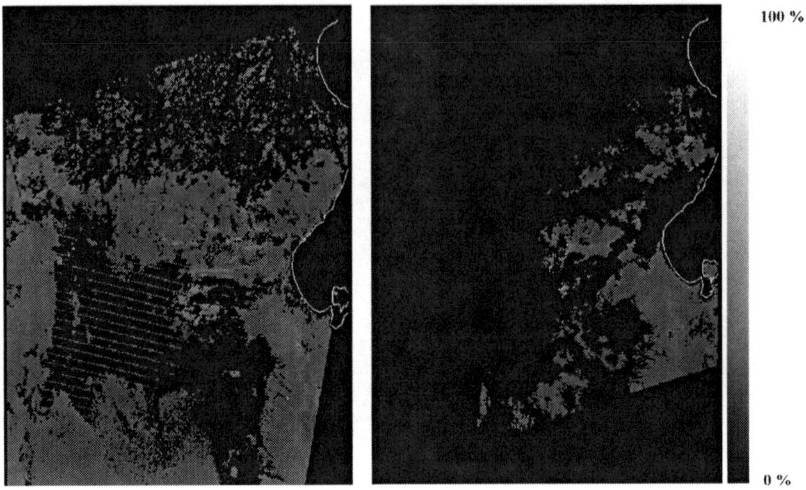

Fig. 6: Maps of soil moisture relative to the first (15th January 2007; right) and second (2nd May 2007; left) sampling events as derived from MODIS data

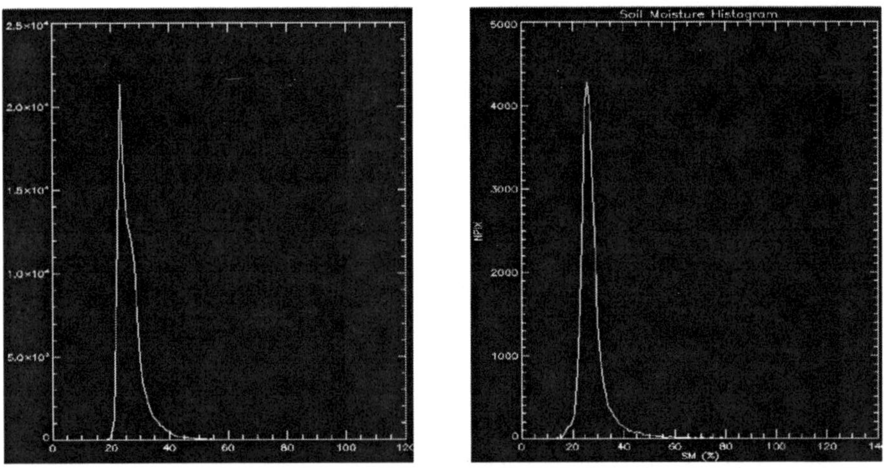

Fig. 7: Frequency distribution of soil moisture relative to the first (15th January 2007; right) and second (2nd May 2007; left) sampling events as derived from MODIS data

It was not possible to compare ground-based measurements with MODIS data since all collected measurements fell into a single MODIS pixel.

3.2 ASAR data

Figure 8 shows maps of backscattering coefficients σ obtained from the two different polarizations.

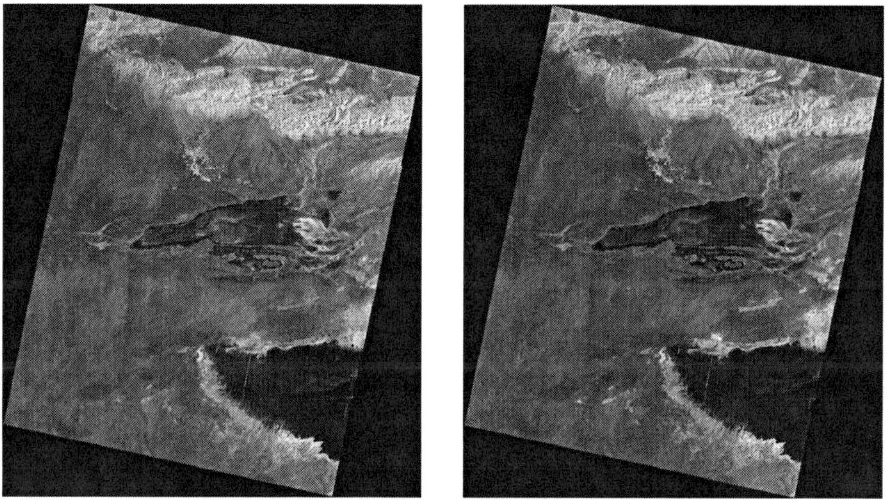

Fig. 8: Maps of backscattering coefficients (Left: σ_{VV}; right: σ_{HH})

When backscattering data were correlated with soil moisture (as estimated from MODIS data and remapped to a common grid, see Fig. 9), no apparent relation emerged (Fig. 10).

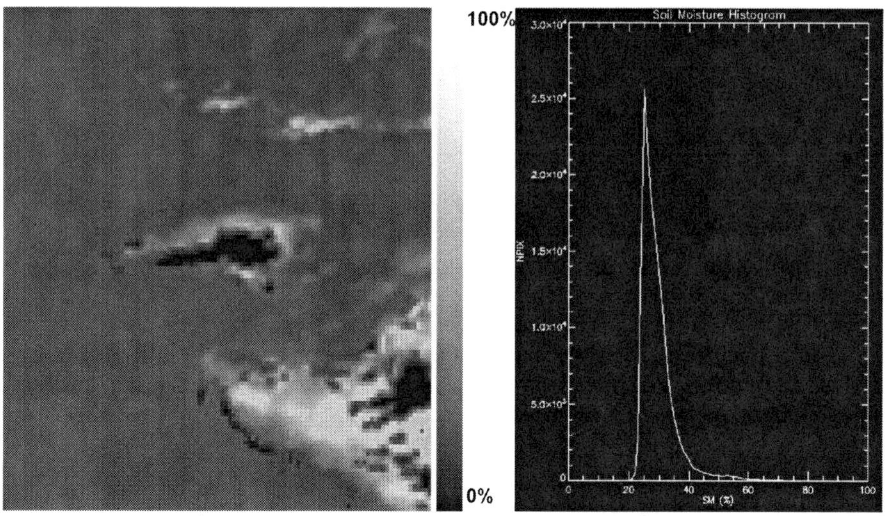

a) b)

Fig. 9: First event MODIS-based SM map remapped over the ASAR domain (a) and relative frequency distribution histogram (b)

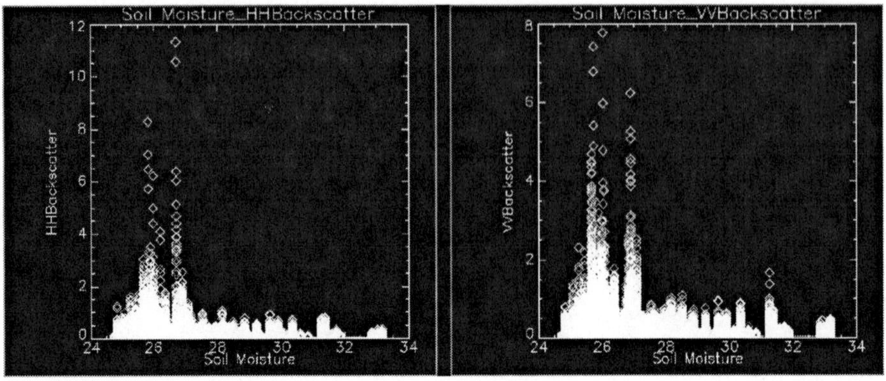

Fig. 10: Relation between backscattering coefficients and soil moisture as estimated from MODIS data (Left: σ_{HH}; right: σ_{VV})

Similarly, the ratio between the two backscattering coefficients showed no apparent relation with soil moisture as estimated from MODIS data (Fig. 11).

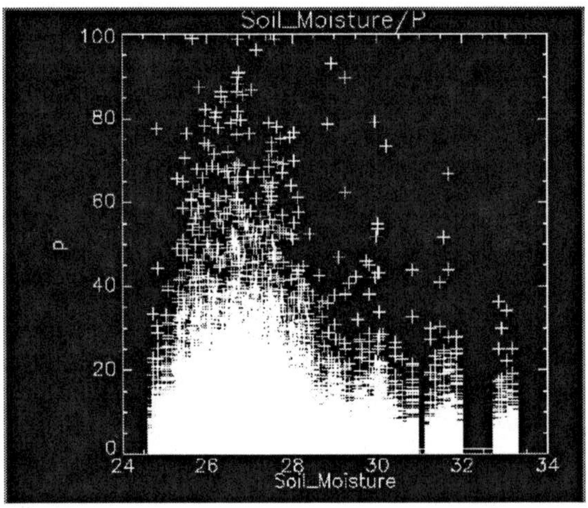

Fig. 11: Relation between the ratio of backscattering coefficients ($p = \sigma_{HH}/\sigma_{VV}$) and soil moisture as estimated from MODIS data

Table 1 and Fig. 12 show backscattering data in relation to soil moisture as derived from direct ground measurements. Again, no apparent relationship emerged, although the sample size is rather small.

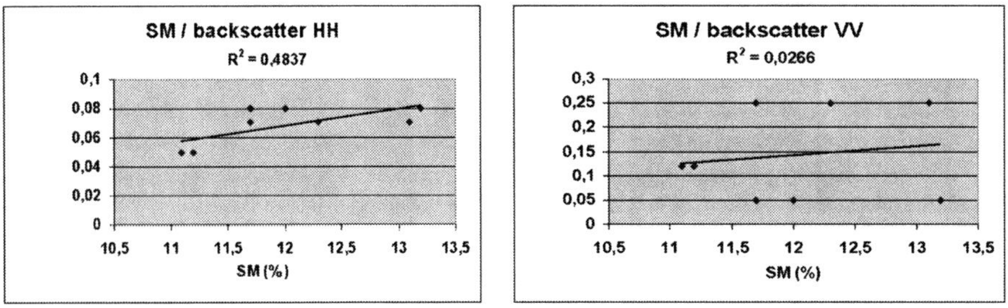

Fig. 12: Relation between backscattering coefficients and soil moisture as measured on the ground (Left: σ_{HH}; right: σ_{VV})

4 Discussion

The availability of a tool for the real time assessment of soil humidity would have important applications, from hydrological monitoring and management, to the improvement of agricultural practices. In this paper, we have reported the preliminary results of an effort directed to developing such a tool.

Estimates of soil moisture obtained from remotely sensed data were consistent with known values of soil moisture commonly recorded in the study areas. Also, a decrease in the estimated average soil moisture from January to May is consistent with the expected seasonal trend. Beside these qualitative tests, however, we were unable to rigorously verify the fit of our satellite estimates with actual soil moisture due to the unavailability of sufficient ground data.

Thus, while this study hints at the possibility of estimating soil moisture over large areas on the basis of satellite passive sensors providing data on land surface temperatures, the actual reliability of this method remains unknown. Further studies including a greater effort in obtaining ground data will be required in order to refine the method.

Bibliography

Alvarez-Mozos J., Casali, J., Gonzalez-Audicana M., Verhoest N. E. C. 2006. *Assessment of the operational applicability of RADARSAT-1 data for surface soil moisture estimation*, in IEEE Transactions on Geoscience and Remote Sensing, n.**44** (4): 913–924.

Baghdadi N., Zribi M. 2006. *Evaluation of radar backscatter models IEM, OH and Dubois using experimental observations*, in International Journal of Remote Sensing, n.**27** (18): 20 September 2006, 3831–3852.

Bignami C., Pierdicca N., Pulvirenti L., Ticconi F. *Risultati preliminari di campagne sperimentali sulla sensibilità dei dati Envisat-ASAR all'umidità del terreno*, XV Riunione Nazionale di Elettromagnetismo RINEM2004, Cagliari, http://www.elettromagnetismo.it/atti_rinem/2004S14A05.pdf

Boisvert J. B., Crevier Y., Pultz T. J. 1996. *Regional estimation of soil moisture using remote sensing*, in Canadian Journal of Soil Science, n.**76** (3): 325–334.

Cai G., Wu J., Xue J., Hu Y., Guo J., Tang J. 2005. *Soil Moisture Retrieval from MODIS data* in Northern China Plain Using Thermal Inertia Model (SoA-TI), 0-7803-9050-4/05 2005 *IEEE.*

Cravey R. L., Jackson T. J., Hsu Ann Y. 1998. *ERS-2 SAR Backscattering Coefficient and Soil Moisture for the Southern Great Plains 1997 Hydrology Experiment.* Retrieval of Bio- and Geo-Physical Parameters from SAR Data for Land Applications Workshop, ESTEC, 21–23, October 1998.

Dubois P.C., Vanzyl J., Engman T. 1995. *Measuring soil-moisture with imaging radars*, in IEEE Transactions on Geoscience and Remote Sensing, n.**33** (4): 915–926.

Glenn, N. F., Carr J. R. 2004. *Establishing a relationship between soil moisture and RADARSAT-1 SAR data obtained over the Great Basin, Nevada, USA*, in Canadian Journal of Remote Sensing, n.**30** (2): 176–181.

Liang Shunlin 2000. *Narrowband to broadband conversion of land surface albedo I algorithms*, in Remote Sensing of Environment 2000, n.**76**, 213–238

Ma A., Xue Y. 1990. *A study of remote sensing: information model of soil moisture*, in Proceedings of the 11th Asian Conference on Remote Sensing, n.**1**, 11.1–11.5.

MEEO *SOIL MAPPER TM Report description document* 2007. Issue 3.7, Date: 12/03/2007, http://www.meeo.it/docs/SOIL_MAPPER_report.pdf

Munro R. K., Lyons W. F., Shao Y., Wood M. S., Hood L. M., Leslie L. M. 1998. *Modelling land surface–atmosphere interactions over the Australian continent with an emphasis on the role of soil moisture.* Environmental Modelling and Software, n.**13** (3–4): 333–339.

Paloscia S., Pampaloni P., Pettinato S., Poggi P., Santi E. 2005. *The retrieval of soil moisture from Envisat-ASAR data*, in European Association of Remote Sensing Laboratories – EARSeL eProceedings, n.**4** (1): 44–52, http://www.eproceedings.org/static/vol04_1/04_1_paloscia1.html

Portmann F. 2000. *The Land-SAF Soil Moisture Product of MIUB and BfG*. Paper presented at the CM-SAF (EUMETSAT Climate Monitoring Satellite Application Facility) Training Workshop, Dresden, Germany, 20–22 November 2000, pp. 6

Tucker C. J. 1979. *Red and photographic infrared linear combinations for monitoring vegetation*, in Remote Sensing of Environment, n.**8** (2): 127–150.

Turner D. P., Ritts W. D., Cohen W. B., Maeirsperger T. K., Gower S., Kirschbaum A. A., Running S. W., Zhao M., Wofsy S. C., Dunn A. L., Law B. E., Campbell J. L., Oechel W. C., Kwon H. J., Meyers T. P., Small E. E., Kurc S. A., Gamon J. A. 2005. *Site-level evaluation of satellite-based global terrestrial gross primary production and net primary production monitoring*, in Global Change Biology, n.**11**: 666–684.

Verstraete M. M., Pinty B. 1996. *Designing optimal spectral indexes for remote sensing applications*, in IEEE Transactions on Geoscience and Remote Sensing, n.**34** (5): 1254–1265.

Wan Z. 1999. *MODIS Land-Surface Temperature Algorithm Theoretical Basis Document (LST ATBD).* Institute for Computational Earth System Science University of California, Santa Barbara, Version 3.3, April 1999.

Xue Y., Cracknell A. P. 1995. *Advanced thermal inertia modelling*, in International Journal of Remote Sensing, n.**16** (3): 431–446.

MONITORING SENSITIVE AREAS TO DESERTIFICATION IN SARDINIA: THE CONTRIBUTE OF THE REGIONAL AGROMETEOROLOGICAL SERVICE

ANDREA MOTRONI[1], SIMONA CANU[1], GIUSEPPE BIANCO[1], AND GIOSUÈ LOJ[2]
[1]*SAR – Regional Weather Service, viale Porto Torres 119, Sassari, Italy, motroni@sar.sardegna.it*
[2]*LAORE – Tecnica in Agriculture, via Caprera 8, Cagliari, Italy, gloj@tiscali.it*

Abstract

It has been almost 10 years since the Agro-meteorological Service of Sardinia has been involved in studying and monitoring desertification phenomena (Bianco et al., 2000). Desertification is meant to be a dynamic process, temporally distributed, which negatively influences ecosystems, diminishing productivity potential of natural resources. The main causes can be identified in climate change and human impacts, as stressed by the definition of desertification given by United Nations Environmental Program ("land degradation in arid, semi-arid and dry sub-humid areas resulting from various factors, including climatic variations and human activities", UNEP, 1994) and commonly adopted by the United Nations Convention to Combat Desertification (UNCCD).

Research activities have been concentrated mostly on mapping the areas sensitive to desertification in Sardinia and investigating particularly the climatic indexes involved (mainly drought and aridity). A Geographic Information System was therefore set up and several thematic layers related to all different indicators obtained. The study was carried out on 16 areas of Sardinia at 1:100,000 scale in collaboration with Sardinia Regional Extension Service. It was also implemented in the whole island territory at a 1:250,000 scale as a part of the National Action Plan against drought and desertification. So far it was used in the National Atlas of Desertification (Costantini et al., 2007) and in the Sardinia Water Protection Plan.

Keywords: Sardinia, Desertification, ESAs methodology, GIS

1 Materials and methods

The Environmentally Sensitive Areas (ESA) methodology, developed in the EU project MEDALUS (Mediterranean Desertification and Land Use) by Kosmas et al. (1999a) was applied to the following 1:100,000 areas (Scheme 1) where field surveys were also conducted:

A. Marini and M. Talbi (eds.), *Desertification and Risk Analysis Using High and Medium Resolution Satellite Data*,

ESAs methodology is the most diffused one and results can be easily compared with those achieved in similar studies in other areas of the world based on the same methodology. Also, there is a good integration of physical, biological and human factors involved in desertification processes. Fifteen bio-physical and social-economic indicators and four indices which classify the territory in potential, fragile and critical areas to desertification were considered.

A scheme of the ESAs methodology is shown in Scheme 2:

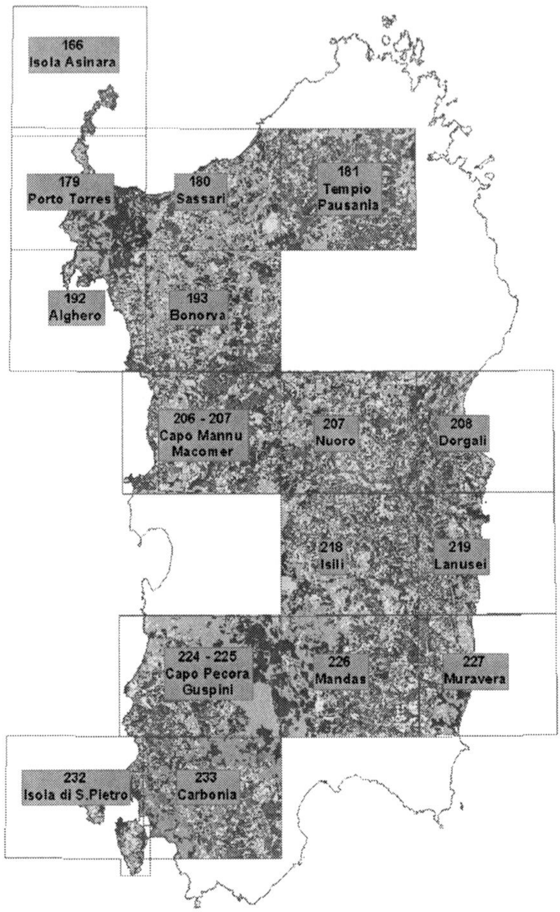

Scheme 1. Scheme of 1:100,000 scale sensitive areas to desertification according to ESAs methodology

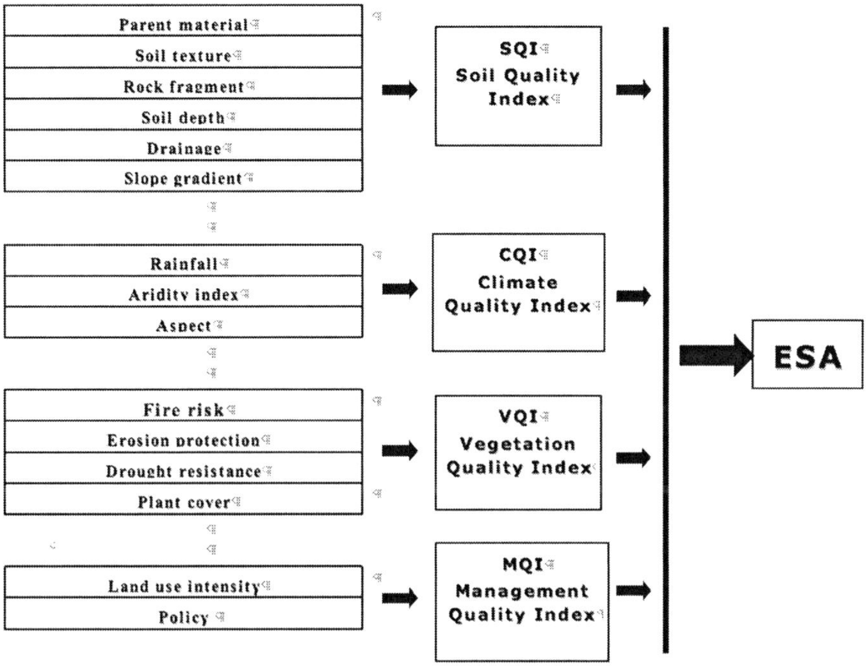

Scheme 2. The general scheme of ESAs methodology

The Quality indexes and the related indicators are here briefly described.

1.1 Soil Quality Index (SQI)

(SQI) is obtained, using GIS techniques, by the geometric means of the following six indicators:

- *Parent material*: soil properties (erodibility and water retention capacity) depend on physical and chemical characteristics of original parent material (D'Antonio et al., 1999).
- *Soil texture*: the typical USDA classification has been used (sandy, clay, loamy and all interactions among these types considered).
- *Rock fragment*: soil erosion and run off are greater in areas characterized by high rock fragments percentage (up to the 20% threshold) due to intense and prolonged precipitation events (Bunte et al., 1993). Beyond this value, the limited space between fragments prevents the development of scour holes and thus limits soil loss.
- *Soil depth*: a higher plant cover characterizes deep soil because plants can find greater water storage and therefore they present a greater resistance to erosion and desertification processes.
- *Drainage*: it is important in soil salinization processes. Saline transport and concentration in soil profile reflect prevailing water balance conditions and the soil depth to groundwater (Cara et al., 1994; Indiati et al., 1998).

- *Slope gradient*: slope classes computation has been completed through the application of a Digital Elevation Model (DEM) with a 20 m resolution.

1.2 Climate Quality Index (CQI)

(CQI) is a key point for detecting how Mediterranean climate influences desertification processes (Motroni et al., 2005). CQI is obtained, using GIS techniques, by the geometric means of the following three indicators.

- *Rainfall:* atmospheric conditions characterizing desert climate lead to severe water deficit, i.e. potential evapotranspiration (ETo) values higher than precipitation values. Such conditions are calculated by several indexes, the most used one is the FAO-UNESCO bioclimatic index (1989), P/ETo.

  (a) Arid zones — $P/ETo < 0.3$
  (b) Semi-arid zones — $0.3 < P/ETo < 0.5$
  (c) Sub-humid zones — $0.5 < P/ETo < 0.75$

Both means climatic rainfall amount and ETo values were estimated for the reference period 1971–2000. Daily data from about 200 meteorological stations were considered. Interpolation of data in a regular grid of 240 cells path 10 × 10 km was then estimated.

- *Aridity index*: originally, ESAs methodology considered the Bagnouls-Gaussen aridity index:

$$BGI = \sum_{i=1}^{n} (2Ti - Pi)\,k$$

BGI = Bagnouls-Gaussen Index
Ti = temperature of the i month (°C)
Pi = total monthly precipitation of the i month (mm)
K = frequency of the condition 2Ti-Pi > 0 for the i month (%)

In this way, the soil component is not considered and much information about water actually available in soil for ecosystems sustainability is lost. For this reason, a simple daily soil water balance model, proposed by Reed et al. (1997), was partially modified and then implemented on climate data and soil properties of Sardinia in order to calculate the aridity index (Fiori et al., 2003). For each year, aridity index values have been estimated computing the number of days in which soil humidity values were below different thresholds of AWC (0%, 10%, 25%, 50%, 75%). The 50% threshold was used for calculating the aridity index in order to avoid over-estimates and underestimates of the index and to obtain a good spatial variability.

Pedological information of Sardinian soils was used to define the entire territory of the island in terms of available water content (AWC) and soil depth. Reference evapotranspiration values (ETo) were estimated using the Hargreaves method and actual evapotranspiration estimates (ETa) were calculated as a function of soil water content. A simplified soil water balance was calculated for each grid point and year to determine the average daily soil water content over the 30-year period 1971–2000 using the following equation:

$$w_i = w_{i-1} + P - f_{i-1}\, ETo$$

w_i = current soil moisture for the i day
w_{i-1} = soil moisture in the previous day
P = precipitation
ETo = potential evapotranspiration
f = evapotranspiration coefficient
f_{i-1} = w_{i-1}/w^*= evapotranspiration coefficient for the day i-1
w^* = Available Water Capacity (AWC)

The aridity index was estimated counting the number of days when soil water content was below the 50% of available water content.

- *Aspect:* Slope aspect is considered an important factor for land degradation processes. Aspect affects the microclimate by regulating the angle and duration at which the sun rays strikes the surface of the soil. In Mediterranean region slopes with southern and western facing aspects are warmer and have higher evaporation rates and lower water storage capacity than northern and eastern aspects (Kosmas et al., 1995). Therefore, a slower recovery of vegetation and higher erosion rates are expected in southern and western than in northern and eastern facing aspects. The basic cartographic information is a Digital Elevation Model (DEM) with a spatial resolution of 20 m, which gives a good detailed view of the territory and make it possible to consider the North and South aspect. Being the information too detailed in those areas where the slope is not enough to individuate the aspects, only the aspect of the areas with a slope >5% have to be considered.

1.3 Vegetation Quality Index (VQI)

Vegetation is important in desertification processes because it is able to reduce soil loss due to rain impact and surface runoff. This is particularly true in hilly areas under arid conditions with year rainfall mean values less than 300 mm and high evapotranspiration rates (Clark et al., 1996). In such conditions, vegetation is very poor and it is not capable of maintaining soil *in situ* avoiding runoff (Francis et al., 1990).

CORINE Land Cover map (scale 1:100,000) has been used as basic layer for defining vegetation quality of Sardinian territory because commonly used in most of EU countries (Delogu et al., 1988; Loj et al., 1998). In ESAs methodology only 12 classes of vegetation are considered (see Table 1), so a more detailed land use map than CORINE may be worthless (Kosmas et al., 1999b).

Vegetation Quality Index is obtained, using GIS techniques, by the geometric mean of the following four indicators:

- *Fire risk.* In our environment, ecosystems characterized by species of Mediterranean maquis and pine forests are at a bigger risk.
- *Soil erosion protection.* Mixed Mediterranean maquis with evergreen forests, pine forests, perennial agricultural crops (particularly olives) and permanent grassland are the vegetation classes which best give protection against erosion events.

Table 1: Vegetation type classes considered in ESAs methodology

Class	Vegetation type
1	Mixed Mediterranean maquis /evergreen forest
2	Mediterranean maquis
3	Permanent grassland
4	Annual grassland
5	Deciduous forest
6	Pine forest
7	Evergreen forest except pine forest
8	Evergreen perennial agricultural crops
9	Deciduous perennial agricultural crops
10	Annual winter agricultural crops
11	Annual summer agricultural crops
12	Bare land

- *Plant drought resistance.* Mediterranean maquis species have developed several strategies to resist to drought conditions, while deciduous perennial crops (orchards, almonds), annual grasslands and annual crops (cereals) are not very resistant.
- *Plant cover.* An area or a territory is under desertification processes if biomass production per hectare is under a defined threshold, which corresponds to about 40% of plant cover. Hilly areas can be highly eroded with less than 40% of plant cover.

1.4 Management Quality Index (MQI)

This indicator is obtained, using GIS techniques, by the geometric means of the following two indicators:

- *Land use intensity*. Human pressures and impacts on environment related to desertification are considered. Different situations are evaluated:

Agricultural areas may experience heavy human pressures if incorrect practices are conducted. Conditions to be considered are the type of crop, irrigated or rainfed areas, level of mechanization, use of fertilizers, herbicides and pesticides.

Pasture lands. In a territory like Sardinia, overgrazing can be a major issue in terms of desertification (Aru et al., 1989; Madrau et al., 1999). In this case land use intensity is defined by estimating the Actual Stocking Rate (ASR) and the Sustainable Stocking Rate (SSR).

The Actual Stocking Rate (ASR) was calculated from the statistics about livestock in all municipal administrative territories of the region (ISTAT 2001) and expressed in *Animal Units* per hectare. The Sustainable Stocking Rate (SSR) is calculated in different steps:

a) Pasture land extension has been computed from several CORINE Land Cover classes (rainfed agricultural crops, grassland, mixed agricultural and natural areas, mixed agricultural and forested areas and natural pastures).

For each class of vegetation b) an estimated mean value of dry matter (kg/ha) has been assigned, from very high (irrigated grasslands) to very low values (for example, mixed agricultural and forested areas).

For each municipal administrative territory, the total mean estimated production of dry matter has been calculated for all grazing areas.

Sustainable Stocking Rate (SSR) computation

SSR = Sustainable Stocking Rate $SSR = \frac{SStot \times C.U.}{average\ weight\ A.U. \times L.I. \times year}$

SStot = Total dry matter/year (q)

C.U. = Utilization Coefficient = (mean annual ingested dry matter/ mean annual available dry matter

Average weight A.U. = average weight of A.U. (kg)

L.I. = mean daily Ingestion Level (%)

Year = duration of grazing (days)

Applying the equation to A.U. $SSR = \frac{SStot \times 0.7}{500kg \times 0.5 \times 360} = A.U./ha$

The approximations of this equation are greatly balanced by the values considered in the final index, where a high use intensity is reached when ASR > 1.5*SSR, i.e. when the actual stocking rate is higher than the sustainable one by 1.5 times.

Natural areas. In the methodology, natural areas are considered those with spontaneous vegetation typical of Mediterranean basin (Mediterranean maquis/ evergreen forests, Mediterranean maquis, pine forests, garigues.

- *Protection policies.* This indicator takes into account any protection policies of the environment that can control the impact of human activities in soil degradation and desertification. Polices related to environmental protection are classified according to the degree to which they are enforced. The information on the existing policies is collected and then the degree of implementation/enforcement is evaluated.

Main protected areas are National Parks, Regional Parks (L.R. n. 31/89), areas with Hydrogeological constraints (R.D.L. 3267/23), Forest conservation areas and SIC sites BIOITALY (EU dir. 92/43). Policy enforcement is evaluated in each case.

2 Estimate of final Environmentally Sensitive Areas Index (ESAI) to desertification

The final results of the methodology is the ESA Index which is obtained by the geometric means of the four indexes (SQI, CQI, VQI and MQI). GIS technology was applied for calculating as thematic layers the geometric means as follow:

$$ESAI = (SQI*CQI*VQI*MQI)^{1/4}$$

ESAs final index indicates areas with increasing sensitivity to desertification from areas not affected to very critical conditions, where land degradation is already ongoing.

Table 2: Different ESAs classes with increasing sensitivity to desertification and unclassified areas

Type	Description
Not affected	Areas with deep to very deep, nearly flat, well drained, coarse-textured or finer soils, under semi-arid or wetter climatic conditions.
Potential	Areas threatened by desertification under significant climate change, if a particular combination of land use is implemented or where offsite impacts will produce severe problems elsewhere.
Fragile	Areas in which any change in the delicate balance between natural and human activity is likely to bring about desertification.
Critical	Areas already highly degraded through past misuse, presenting a threat to the environment of the surrounding areas.
Unclassified	Bare rocks, water bodies, urban areas.

3 Main results achieved

ESAs methodology showed its versatile capability to determine and identify the most sensitive areas to desertification and main territorial weaknesses at a local scale (Pittalis, 2003). This study was achieved by the calibration and implementation of several indicators and indexes related to soil, climate, vegetation characteristics and to the management of the territory.

Similar studies have been accomplished during the last years in several Italian Administrative Regions (Sicilia, Calabria, Puglia, Abruzzo, Campania and Basilicata) and also in other European countries (in Spain for example, in Andalusia and Murcia Regions).

In the present study, some adaptations have been made to the original methodology in order to cope with Sardinian specific environmental characteristics and constraints. In particular:

CORINE Land Cover classes were accurately converted to the ESAs vegetation classes.

Slope and aspect classes were remodelled with a minimum threshold.

Climate Quality Index was more deeply studied and investigated, and proper spatialization techniques were implemented. A particular effort has been put in estimating an aridity index based on a simplified water balance, which considers also the soil characteristics, rather than a simply atmospheric balance.

For pasture lands, an overgrazing risk indicator was calculated based on the animal units method. Even if it is still roughly estimated, this index can identify areas were the animal stock is by far not sustainable.

Protection policies considered in this work are only those efficiently applied.

In Fig. 1 the final map of ESAs for Sardinia is shown. The most critical areas (dark grey) can be found along the eastern cost of the island, due to the influence of climatic factors, and in the main plains of Nurra and Campidano, where intensive agriculture is concentrated and human pressure on environment is more forceful.

Charts in Figs. 2 and 3 present ESAs classes distribution. As final results, desertification processes critically threaten about 50% of Sardinian territory (about 24,000 km^2), and about 11% of the island is somehow already highly degraded.

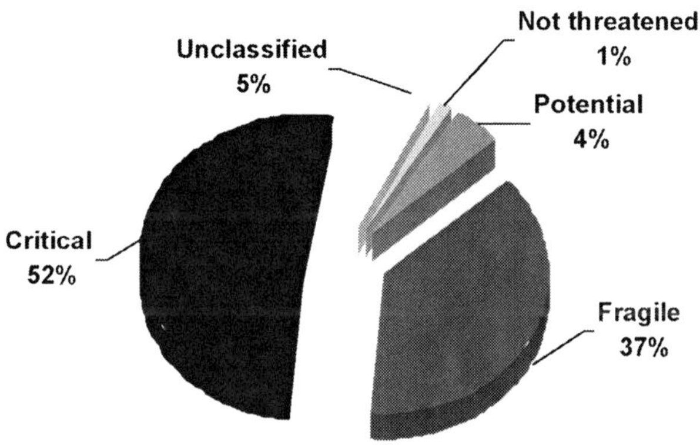

Fig. 1: Percentage of Sardinia total territory in the ESAs classes

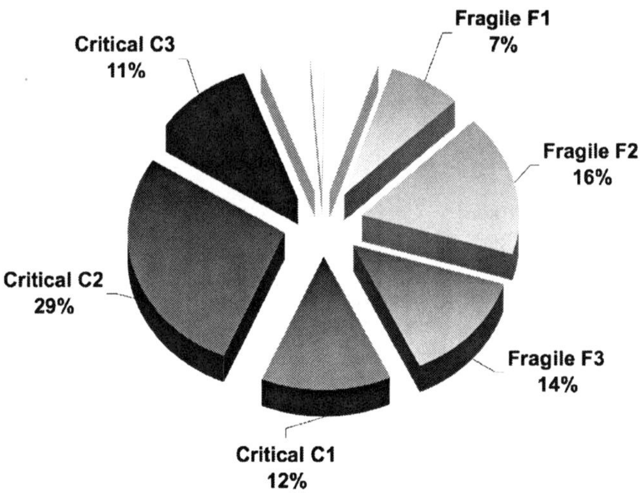

Fig. 2: Subdivision of ESAs critical and fragile classes as percentage of Sardinian territory

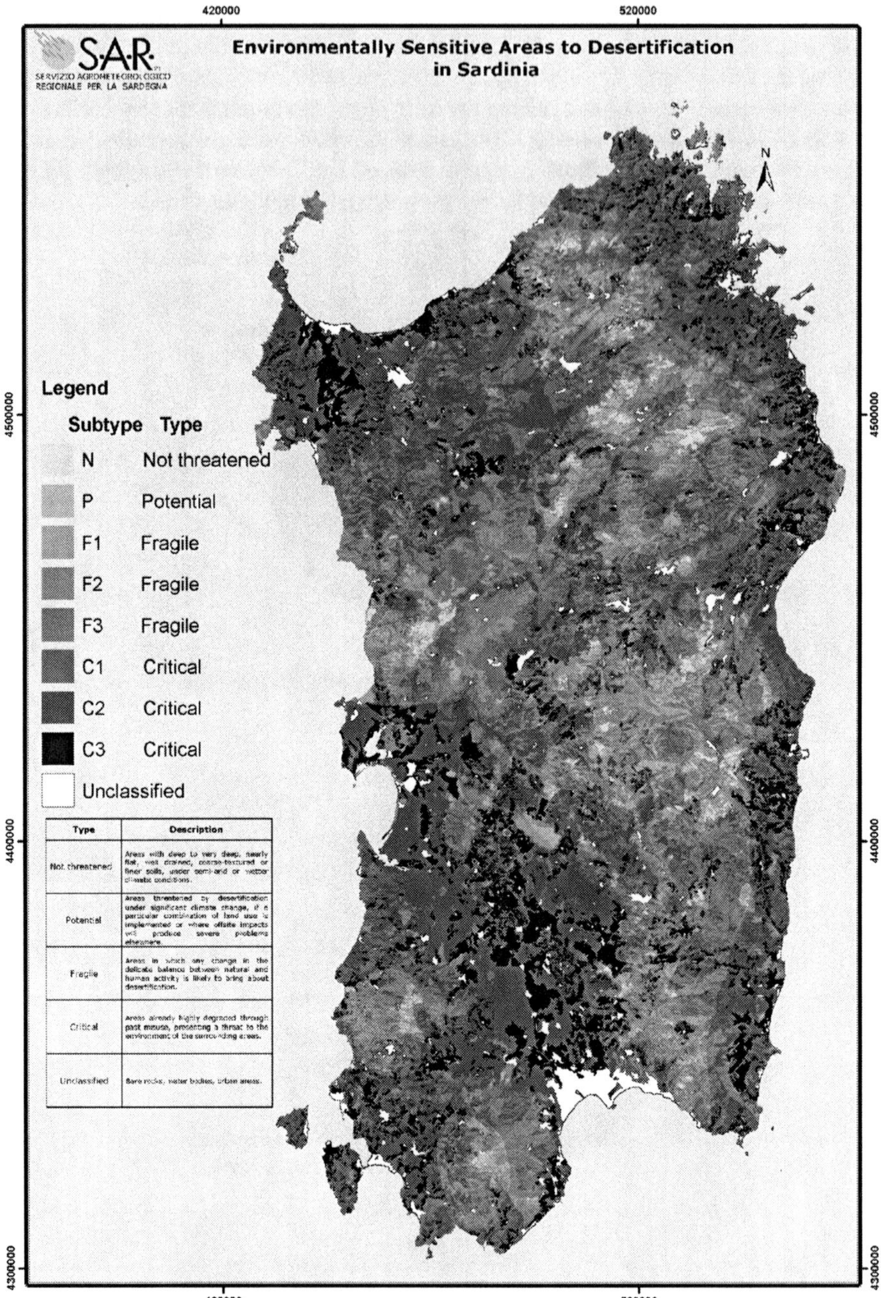

Fig. 3: The environmentally sensitive areas to desertification of Sardinia

Bibliography

Aru A., Baldaccini P., Loj G. 1989. *I suoli: caratteristiche che determinano la marginalità e la loro valutazione per il pascolo.* Atti del C.N.R. Progetto I.P.R.A.

Bianco G., Loj G. 2000. *Il Programma per la lotta alla siccità e alla desertificazione della Regione Sardegna nel contesto nazionale: il contributo dell'ERSAT e del SAR.* Convegno regionale su: siccita', monitoraggio, mitigazione, effetti. Villasimius 21–23 settembre 2000.

Bunte K., Poesen J. 1993. *Effects of rock fragment covers on erosion and transport of non-cohesive sediment by shallow overland flow.* Water Resources Research n. **29**:1415–1424.

Cara S., Demontis A., Di Gregorio F., Loj G., Marini A., Pistis. 1994. *Metodologia d'indagine sul dissesto idrogeologico e sull'erosione del suolo in aree compromesse: il settore di Ingurtosu.* Poster di presentazione dei lavori per il progetto Ambiente, finanziamento R.A.S-Università di Cagliari.

Clark S.C. 1996. *Mediterranean ecology and an ecological synthesis of the field sites*, in: J. Brandt and J. Thornes (eds), Mediterranean desertification and land use, Wiley, London, 271–299 pp.

Costantini E.A.C., Urbano F., Bonati G., Nino P., Fais A.(curatori) 2007. *Atlante nazionale delle aree a rischio di desertificazione.* INEA, Roma, pp. 108

D'Antonio A., Tosco D., Aramini G., Bellino F., Guaitoli F., Di Lisa G.A., Loj G., Pumo A., Reale T., Viviano L. 1999. *Aspetti della sensibilità e vulnerabilità dei suoli in Italia meridionale e insulare*, Firenze 18–20 ottobre 1999.

Delogu G., Loj G., Marquetti R., Marini A. 1988. *Dati landsat TM: confronto di metodologie per l'elaborazione di indici di vegetazione in Sardegna.* Atti del II Convegno A.I.T.1988, Bolzano.

FAO-UNESCO 1989. *International soil reference and information centre. Soil map of the world. Revised legend*, in World Soil Resources Report n. **60**. Roma.

Fiori M., Motroni A., Duce P., Spano D. 2003. *A daily water balance estimate for climate risk evaluation at a local scale.* Proceedings of the Fourth International Symposium on Irrigation of Horticultural Crops – University of California, Davis 1–6 September 2003, Acta Horticulturae n. **664**: 237–243.

Francis C.F., Thornes J.B. 1990. *Runoff from three Mediterranean vegetation cover types*, in Vegetation and erosion, processes and environments, Wiley, Chichester, 333–352 pp.

Indiati R., Loj G. 1998. *Use of alternative soil phosphorus test for improving environmental management of soil phosphate.* Rivista di agricoltura mediterranea, n. **128**, 298–306.

Kosmas C., Moustakas N., Danalatos N.G., Yassoglou N. 1995. *The effect of land use change on soil properties and erosion along a catena*, in: J. Brandt and J. Thornes (eds), Mediterranean desertification and land use, Wiley, 271–299 pp.

Kosmas C., Kirkby M., Geeson N. 1999a. *Manual on key indicators of desertification and mapping environmentally sensitive areas to desertification*, in the MEDALUS project – Mediterranean desertification and land use. European Commission, Brussels.

Kosmas C., Poesen J., Briassouli H. 1999b. *Key indicators of desertification at the ESA scale*, in: the MEDALUS project – Mediterranean desertification and land use. European Commission, Brussels.

Loj G., Scampuddu S. 1998. *Sperimentazione della legenda CORINE-LAND COVER in scala 1:25.000, in due aree campione della Sardegna.* Convegno di presentazione del progetto CORINE, Rivista del Centro Interregionale di Cartografia, Bologna 20 Ottobre 1998.

Madrau S., Loj G., Baldaccini P. 1999. *A model for evaluating the suitability of grazing land to the sardinian soils.* Internal pubblication of ERSAT and MEDALUS project, Cagliari. (Modello per la valutazione dell'attitudine al miglioramento dei pascoli dei suoli della Sardegna. ERSAT-MEDALUS, Sassari, Stampa in proprio.)

Motroni A., Canu S. 2005. *Climate indicators for assessing sensitive areas to drought and desertification in Sardinia (Italy).* WMO/FAO/Cost Action Workshop on "Climatic analysis and mapping for agriculture". Bologna, 14–17 June 2005. In press.

Pittalis D. 2003. Applicazione di una metodologia per l'individuazione di aree sensibili alla desertificazione nel territorio comunale di Sassari mediante elaborazione GIS. Tesi di laurea, A.A. 2001–2002. Università degli Studi di Sassari – Facoltà di Agraria.

Reed S.M., Maidment D.R., Patoux J. 1997. *Spatial Water Balance of Texas*. CRWR Online Report 97-1.

CONTRIBUTION OF SAR INTERFEROMETRY (FROM ERS1/2) IN THE STUDY OF AEOLIAN TRANSPORT PROCESSES: THE CASES OF NIGER, MAURITANIA AND MOROCCO

CATHERINE BODART[1,2], JEAN GASSANI[3], MARC SALMON[2], AND ANDRÉ OZER[2]
[1]*FNRS Research Fellow - Ph.D. Student*
cbodart@ulg.ac.be
[2]*Geomorphology and Remote Sensing Laboratory, Unit of Physical Geography, University of Liège*
BAT. B11, allée du 6 Août, 2, B-4000 Liège, Belgium
aozer@ulg.ac.be; msalmon@ulg.ac.be
[3]*UMR-MA 105 « Landscapes and Biodiversity »*
2 boulevard Lavoisier, 49045 Angers Cedex O1, France
gassani@hotmail.com

Abstract

In order to face desertification and sand encroachment of inter dunes depressions, as well as to protect soils in the area of Gouré (Southeast Niger) where it is possible to have food plantations, it is essential to localize the origin and drift of sands. This paper shows that radar satellite remote sensing and interferometry offers the opportunity to gather all the required information. Digital Elevation Models (DEM) generated by interferometry on ERS 1-2 SAR images (a tandem pair acquired on April 25 and 26 1996) helped to collect information on relief and provided an aeolian model, while giving some insight into the spatial distribution of sand formations. Interpreting the interferometric coherent image by means of a color composite allowed detecting and locating sand movements in the areas under study. This project mainly highlights the contribution of SAR interferometry in the study of sand movements by developing added-value products. Interpretations were validated using high-resolution satellite data (SPOT 5, IKONOS, and QUICKBIRD) and field observations made during a campaign in Niger in March 2004. The same procedure was also applied in Mauritania using two tandem pairs (2–3 November 1995 and 7–8 December 1995) and in Southern Morocco (18–19 May 1996) in order to identify active dunes in these regions. These observations will be taken into account when analyzing ENVISAT-ASAR data.

Keywords: SAR Interferometry, coherence interpretation, active dunes, Sahel

A. Marini and M. Talbi (eds.), *Desertification and Risk Analysis Using High and Medium Resolution Satellite Data*,

1 Introduction

Desertification is considered one of the most critically important issues in many countries (Reynolds et al., 2002). The United Nations Convention to Combat Desertification (UNCCD) defines it as "land degradation in arid, semi-arid and dry subhumid areas resulting from various factors, including climatic variations and human activities" (http://www.unccd.int).

In particular in West Africa desertification is a problem of major importance, which creates a situation of misery together with drought and poverty. In the late 1960s, the whole Sahelian area, like Southern Niger (Fig. 1), was affected by a terrible drought (Ozer, 2000; Ozer et al., 2005). This climatic deterioration during the 1970s and 1980s has generated, together with human factors, serious environmental alterations. Indeed, these regions are now suffering startling land degradation. Following the reduction of the vegetation cover, many dunes which came into existence during the ogolian period (18000 y. BP), and which were formerly stabilized by vegetation, are now reactivated (Fig. 2). Currently, on the edge of the Sahara desert, moving dunes threaten the existence of many farmlands, water points, settlements and transport links (Bodart, 2004; Gassani, 2004; Ozer, 2000; Ozer et al., 2002).

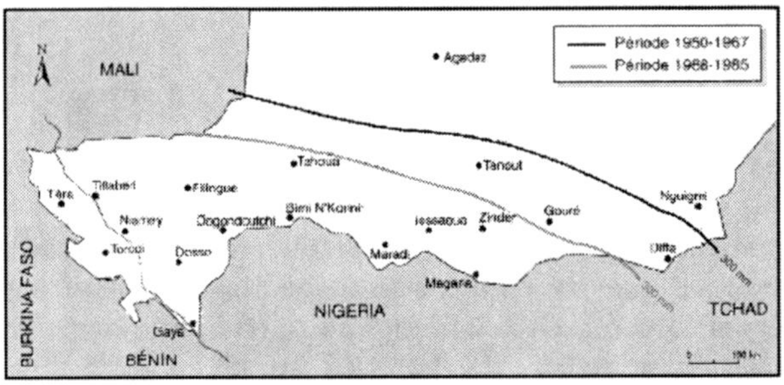

Fig. 1: Pluviometric map of Southern Niger. Migration of the 300 mm isohyet (limit of the mil culture) to the south suggesting an increase in aridity (From Ozer P., 2000)

Fig. 2: Inter dunes depression threatened by a reactivated dune in southeast Niger

In order to fight against sand dunes encroachments and desertification in an efficient manner and ensure successful and sustainable projects, a good understanding of driving mechanisms through an integrated approach including the mapping and monitoring of sensitive zones is required.

Radar, remote sensing and interferometry have proved their utility in a wide range of earth observation studies. The purpose of this paper is to show how SAR interferometry can be used as a tool, among different remote sensing-based techniques, for the study of aeolian transport processes and in particular for the detection of active sand dunes.

Preliminary results of three case studies are presented. The two first areas are located in the Sahelian area: southeast Niger (region of Gouré) and Mauritania, while the last one is located in Southern Morocco (Draa valley) on the northern border of the Sahara desert (Fig. 3).

This study is supported by the ESA ENVISAT program, Project FADA AO824 (Fight Against Desertification in West Africa), by ISIS (*Incitation à l'utilisation Scientifique des Images SPOT*) ISIS0312-571 for the SPOT images and by CUD (Coopération Universitaire pour le Développement, Belgium) and BTC (Belgian Technical Cooperation) for the field trips.

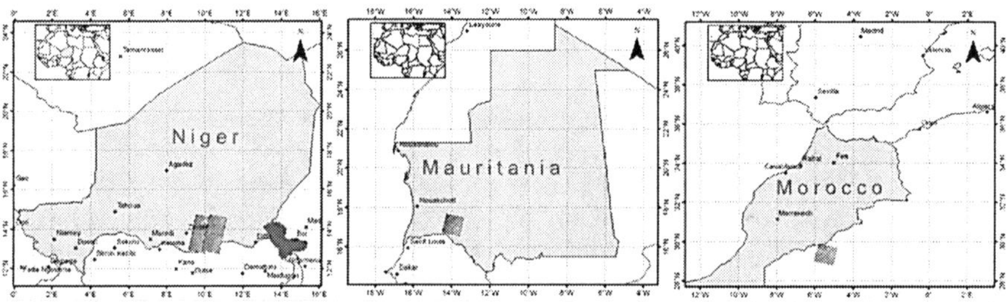

Fig. 3: Location of the three case studies

2 Data set and processing

InSAR processing, including georeference, were performed, using the InSAR/DInSAR processor of the "Centre Spatial de Liège". In a first step, only one-day tandem pairs of interval from ERS 1-2 were acquired (Table 1). Note that only relatively scarce SAR data suitable for interferometry exist in those regions, i.e. on the edge of the Sahara desert, which might prove to be a major limitation for more expanded investigations. In order to study the aeolian formations and transport processes, our evaluation focused on analyzing the following interferometric products:

- DEMs
- Coherence images
- Intensity images

To interpret the coherence and intensity images in an easy manner, RGB color composites were generated with the coherence image in the Red channel, the intensity image of the first acquisition in the Green channel and the intensity image of the second acquisition in the Blue channel.

Table 1: Tandem pairs from ERS 1-2 data for the three case studies

<table>
<tr><th>Region</th><th>Mission</th><th>Date</th><th>Orbit</th><th>Baseline Bp (meters)</th><th>Track</th><th>Frame</th></tr>
<tr><td rowspan="4">Niger</td><td>ERS1</td><td>25/04/1996</td><td>24988</td><td rowspan="2">-91</td><td rowspan="2">451</td><td rowspan="2">3321 + 3339</td></tr>
<tr><td>ERS2</td><td>26/04/1996</td><td>5315</td></tr>
<tr><td>ERS1</td><td>9/04/1996</td><td>24759</td><td rowspan="2">-98</td><td rowspan="2">222</td><td rowspan="2">3321 + 3339</td></tr>
<tr><td>ERS2</td><td>10/04/1996</td><td>5086</td></tr>
<tr><td rowspan="2">Morocco</td><td>ERS1</td><td>18/05/1996</td><td>25318</td><td rowspan="2">-105</td><td rowspan="2">280</td><td rowspan="2">3015</td></tr>
<tr><td>ERS2</td><td>19/05/1996</td><td>5645</td></tr>
<tr><td rowspan="4">Mauritania</td><td>ERS1</td><td>2/11/1995</td><td>22484</td><td rowspan="2">-183</td><td rowspan="2">452</td><td rowspan="2">3267</td></tr>
<tr><td>ERS2</td><td>3/11/1995</td><td>2811</td></tr>
<tr><td>ERS1</td><td>7/12/1995</td><td>22985</td><td rowspan="2">-240</td><td rowspan="2">452</td><td rowspan="2">3267</td></tr>
<tr><td>ERS2</td><td>8/12/1995</td><td>3312</td></tr>
</table>

Thus, a high coherence level (value approaching 1) arises in red indicating a very good phase correlation between the two acquisitions (high scene stability), whereas dark colors suggest a coherence loss (value approaching 0), which means that targets have moved during the one-day interval. Finally, white spots correspond to areas with a high coherence level and a strong radar signal backscattering during the two days, for example, relief zones.

Although the coherence information is generally used as a quality indicator for DEMs, quality and/or phase noise estimation, it can also be used as an information channel in itself (Derauw, 1999). As such, it provided crucial information on small sand dune movements, i.e. movements that occurred within only 24 h.

The same methodology was applied in the three studied areas and all the interpretations were crosschecked with high-resolution optical data (SPOT 5, IKONOS, and QUICKBIRD), aerial photos and field observations.

3 Results

3.1 Niger case study

In the Gouré area, southeast Niger, the transverse dunes formerly fixed by vegetation are now reactivated by a reduction in the vegetation cover and an increase in wind erosion. Several inter dunes basins, villages and many farmlands are threatened by these processes of sand dune encroachment. These inter dunes basins show evidence of an ancient network of temporary rivers active during wet periods in the Quaternary and constitute the main source of production and agricultural diversification in these areas with reduced potentialities. As mentioned above, in this region of Niger, vegetation cover reduction results from variations in aridity (drought) and is also induced by human activities, such as trampling, overgrazing, unsustainable agricultural practices, collecting wood for fuel, expanding urban areas, etc.

On the one hand, the Digital Elevation Model generated by interferometry provided morphologic information and an aeolian model of the studied area while giving some insight into the spatial distribution of sand formations. On the other hand, the color composite inspection enabled to find out where sand dunes movements occur. Coherence losses were identified especially around the city of Gouré (Fig. 3). In the present case, they correspond to a ring of sand dune reworking linked to the urban area. Moreover, isolated active dunes could also be easily spotted within the concerned area (Fig. 4).

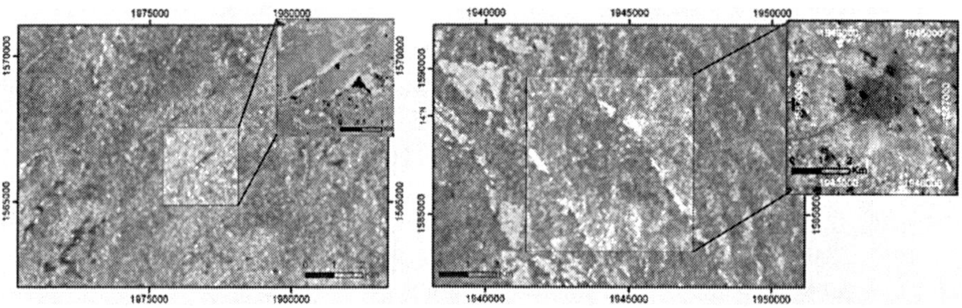

Fig. 4: RGB color composite (tandem pair of 25–26 April 1996). Red: coherence image, green: intensity image of 25April 1996, blue: intensity image of 26 April 1996. Left: isolated active dunes – aerial photo IGN France (1996). Right: city of Gouré – Spot 3 – image (31 October 1996)

3.2 Morocco case study

The Draa valley takes shape, downstream from the city of Ouarzazate, at the junction of the Dades and Ouarzazate wadis. In its middle portion, i.e. from Agdz to M'Hamid, the presence of Ordovician shale allowed the development of wide depressions surrounded by sandstone ridges. These ridges protect six palm groves separated by transverse valleys, named "cluses" or "foums". In this region, rainfalls have a North-South gradient decreasing from 300 to 40 mm per year from upstream to downstream and show strong variations from year to year. In addition to this lack of water, the vegetation is threatened by anthropogenic degradations, such as trampling and overgrazing. As in Niger, these phenomena led to sand dunes encroachment, particularly around the palm groves.

Figure 5 shows the RGB color composite resulting from the tandem pair of 18–19 May 1996 as well as small sections of an IKONOS image and field observations allowing validating the interpretations. The zones with coherence losses were systematically correlated with active dunes or, in general, with processes of sand dune encroachments. In addition, specific areas such as villages, oases or roads were identified in order to help fight desertification in Morocco.

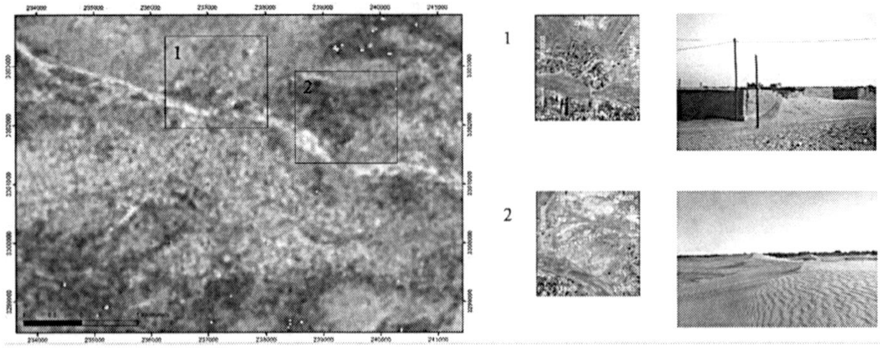

Fig. 5: RGB color composite (tandem pair of 18–19 May 1996). IKONOS image (29 October 2001) – field observations (November 2003), 1: M'Hamid village threatened by sand dune encroachment, 2: active field dune in the east of M'Hamid

3.3 Mauritania case study

The zone of Lake Aleg in southwestern Mauritania, in the northern sahelian climate, is located on the boundary between two very different geo systems. In the north is marked by the presence of longitudinal dunes from the Ogolian period (18000 y. BP) with a Northeast-Southwest orientation. Today these dunes, with a strongly reduced vegetation cover, are subjected to an important reworking of their summits. Large sandstone areas characterize the second geo system in the southeast. During the rainy season, the river system of the Oued Ketchi can be observed.

It takes its source on the plateau of Tagant and drains very large volumes of water down to a depression, i.e. Lake Aleg.

In this region of Mauritania, the activity on the top of the longitudinal dunes was detected clearly and quickly (Fig. 6). To a lesser extend, others coherence losses zones were correlated with water and a specific type of vegetation cover associated with wetlands.

4 Concluding remarks

The validation of the coherence image interpretation using high-resolution data and field observations in the three analyzed regions confirmed that InSAR can be a reliable tool to detect and locate sand movements in the area under concern.

These preliminary results highlight the benefits of radar remote sensing and interferometry through the development of added value products in the study of sand movements and suggest that interferometric radar data can complement other remote sensing data in order to evaluate and indicate desertification in West Africa.

Considering the previous results, a further step of this research will consist in processing and analyzing ENVISAT-ASAR data, as well as in examining the possibility of using InSAR for more robust detection and temporal monitoring of active dunes.

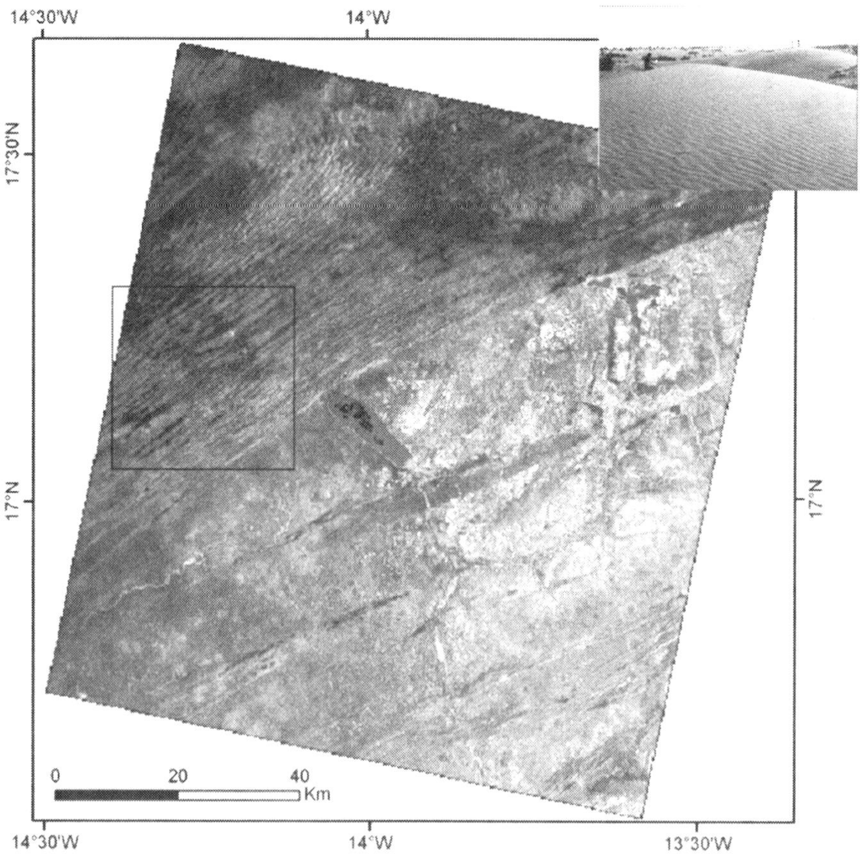

Fig. 6: RGB color composite (tandem pair of 2–3 November 1995) – field observations (October 2003)

Acknowledgments

All the InSAR processing, including georeferencement, were performed thanks to the "Centre Spatial de Liège" InSAR/DInSAR processor. We thank Dr. Dominique Derauw for welcoming us in the laboratory and for helpful discussions.

The authors also thank ESA for ERS SAR data provided under the FADA project.

Bibliography

Bodart C., 2004. *Etude de l'ensablement dans la région de Gouré (sud-est du Niger): processus et apport de la télédétection*, Master Dissertation, Faculté des Sciences, Université de Liège.

Derauw D., 1999. *Phasimétrie par Radar à Synthèse d'Ouverture; théorie et applications*, Ph.D. Dissertation, Faculté des Sciences, Université de Liège.

Gassani J., 2004. *Apport de la télédétection à l'analyse et à la gestion des risques naturels sur la zone du lac d'Aleg* (Brakna, Mauritanie), Master Dissertation, Faculté des Sciences, Université de Liège.

Ozer P., 2000. *Les lithométéores en région sahélienne*, in GEO-ECO-TROP, n.**24**, 1–4.

Ozer A., Legat A., Schyns J.-C., Petit S., Salmon M., Goossens R., De Dapper M., Ost L., 2002. *Etude par télédétection de l'ensablement dans la province de Ouarzazate, Maroc*, Project report, Liège.

Ozer P., Bodart C., Tychon B., 2005. *Analyse climatique de la région de Gouré, Niger oriental: récentes modifications et impacts environnementaux*, Cybergeo, n.**308**.

Reynolds J.F., Stafford Smith D.M., 2002. *The Dahlem desertification paradigm: a new approach to an old problem*, in Global desertification: do human cause deserts? Eds. J.F. Reynolds and M. Stafford Smith. Dalhem University Press, Berlin, Germany, pp. 1–21.

DATA BANK OF TERRITORIAL SPECTRAL SIGNATURES FOR SARDINIA AND NORTH AFRICA

ALBERTO MARINI AND PAOLO COLANDREA
Earth Sciences Department – TeleGIS Laboratory – University of Cagliari, Via Trentino 51, 09127 Cagliari, phone: 070-6757701, Fax: 070-282236
marini@unica.it
paolocolandrea@gmail.com

Abstract

The evolution of the methodologies to use the hyper-spectral images allows a different approach in the data analysis, which previously was difficultly practicable.

Analyzing the general availability of metadata concerning the spectral characterization of physiographic units, still today a remarkable deficiency of acquaintances emerges, which regards the data themselves and the validation systems of the radiometric parameters. The methodologies for data archiving and in particular for the comparison between signatures obtained from different platforms, are still in phase of experimentation, for the difficulties of comparisons between the interpretations, due to the different conditions of the answers in the physic-environmental parameters of the objects to investigate.

This collection of spectral signatures can concur at the integration of the measures detected on the field with satellite data, allowing to set up methodical procedures relative to the hyper-spectral images now available; it will be also possible to develop a collection of signatures, on the earth and over the sea, for all the successive controls.

Keywords: MODIS, Remote Sensing, Database, Infrastructure, Internet, Networks

1 Introduction

The collection of "spectral signatures", through a portable spectrometer MACAM with 1,600 bands from 0.4 to 2.5 μm, allowed to dispose of material currently almost unavailable in other web localizations, useful for a direct quantitative and objective implementing, also for the photo-interpretation phase of remote sensing images.

The data bank of spectral signatures becomes an instrument parallel to the reception system and parallel to the satellite images distribution system organized in Cagliari, near the TeleGIS Laboratory, were a SeaSpace-TeraScan system is continually working, receiving MODIS images in X-band and L-band. This system gives the possibility to receive data from a great number of satellites, as the MODIS data, transmitted from the Terra and Aqua satellites, and images AVHRR from

A. Marini and M. Talbi (eds.), *Desertification and Risk Analysis Using High and Medium Resolution Satellite Data*,

NOAA satellites. A 2.4 m antenna composes the receiving system, protected by a spherical radon that allows operating also in presence of heavy weather conditions. The sensibility also allows the reception of signals from very low passages over the horizon and thus relative to coverage very far from the reception point: the antenna is covering daily from the Canaries Islands to the Black Sea and from Norway to Mali, with resolution, for the MODIS data, in 36 bands with pixels from 250 to 1000 m.

The first two MODIS bands are at medium resolution (250 m). Then there are 5 bands with resolution of 500 m covering the visible and the infrared spectra, and the others, mainly focused on values of the far infrared and the thermal, have a resolution of 1,000 m.

After a first processing step, satellite images are topographically corrected (i.e. geolocated), assigning to each pixel values corresponding to geographic coordinates, which allow images to be represented in all projection/cartographic systems. Data processing over selected areas, allows the generation of geophysical products, based on the IMAPP software of the University of Wisconsin and DACC of NASA. These algorithms permit also the integration of multi-platform data and the elaboration between the same ones, in order to develop new applications deliverable by the web.

The collection of spectral signatures and the data bank of received images use, starting from August 2004, at the moment is based on the connection to the Guardian browser, whose server is located at the TeleGIS. Internet access is ruled by username and password, and a guest user has been created to allow public access (user: telegis; password: telegis).

A Network of Laboratories and Users is under development, to exchange data and results of applications.

2 Characteristics of the receiving system

The TeraScan system of the SeaSpace allows overcoming the double obstacle concerning the data availability and the elaboration of multiple and hyper-spectral images. The antenna has two channels, for signals transmitted in X-band and also in L-band. These data follow two different processes of conversion and however get to the same computer, which manages the antenna with a continuous validation of the ephemeredes of several satellites through Internet.

Several environmental applications are possible by receiving MODIS and AVHRR images, using a sophisticated series of algorithms to analyze various parameters about the atmosphere, the waters and lands.

It is possible to determine profiles of the atmosphere, analysis of the temperature of clouds and appraisal of rains to be exploited by operational meteorology.

The "color" of the sea can be obtained and then compared with maps of daily distribution of temperatures, the indices of phytoplankton and about suspended materials.

On land, biomass variations can be monitored, together with quantification of forest fires and analysis of surface reflection.

All the products are generated by Network partner request and are distributed via Internet.

3 Characteristics of the Guardian platform

Guardian is an Internet client-server GIS platform to manage via TCP/IP a great variety of geographic data, included the satellite images and products, in a GIS environment which allows a great variety of applications, from professional activities of photo interpretation (e.g. agricultural controls, cartography, environmental monitoring, etc.), to the publication on Internet of catalogues of GIS data and satellite images, until support activities for didactics purposes in remote sensing, GIS and environment (Fig. 1).

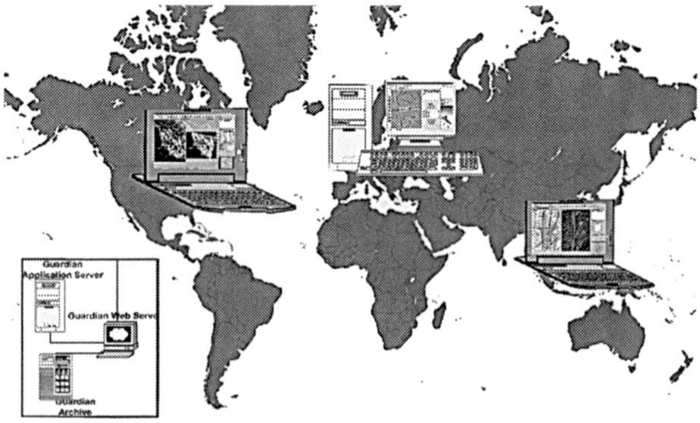

Fig. 1: Outline of operations in the Internet/GIS Guardian platform

The main functionalities of the system are listed below.

- Access with authentication (username and password) to the services via Intranet/Internet (included the download of the client application via Internet)
- Client application multiplatform (Windows, Linux, MacOS, Solaris, etc.)
- Data browsing and selection, raster and vector, from data base
- Centralized elaboration of the data on server side
- RGB composition of selected bands
- Contemporary visualization of images and superimposition of vector data, in up to four panels, with synchronized operations (pan, zoom, etc.)
- Possibility to work with geographic systems and different projections

Guardian is a multiplatform application. The user can utilize it from any computer and the application works for all the most common systems (Windows, Linux, Solaris, MacOS, etc.). The system manages the data in a centralized way loading the data (satellite images, thematic layers, GIS, maps, digital cartography etc.). Therefore every user will be able to browse through the catalogue and to select and to download the necessary data (Fig. 2).

The platform has specific functions for elaboration of centralized images; web users can manipulate the remote execution and visualization directly on the own computer, thanks to the use of a set of images processing functions, also in lack of large-capacity of processing power. All these functions can be used by several users, who may be located anywhere. Guardian can catalogue data processed by single users who can share and approach via Internet their works. The client applications come free of charge, distributed, reducing for the users the costs of the SW licenses and favoring simplicity of access to the data bank of one immense community of users. All the accesses are regulated by user-name and password and the communications are encrypted, guaranteeing the safety of communications.

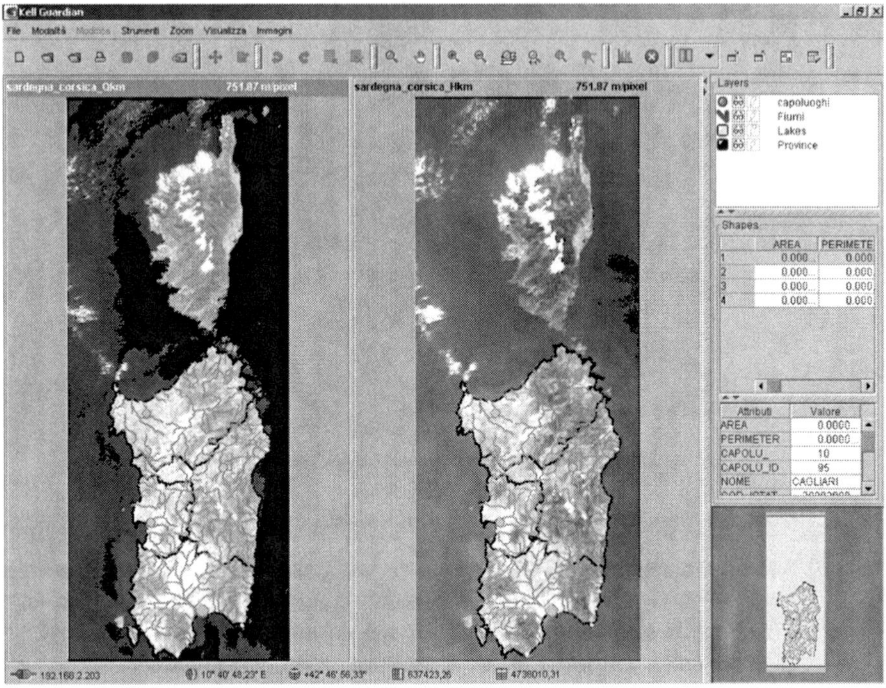

Fig. 2: An example of the visualization of MODIS data and thematic maps, with the client of Guardian

4 Data bank applications

The integration between a continuous, daily series of images from satellite and the relative thematic elaborations, a data bank of spectral signatures and the availability of a on line management system, visualization and crossed analysis of the various types of data also with a vector system, are rendering this new instrument (made available on the Internet by the TeleGIS Laboratory) a precious aid for the analyses of environmental variations in the short and long period, useful to evaluate natural and anthropic action.

Fifteen years of observations is the minimum time to determine the effects and the connections between the several natural phenomena while longer periods are necessary in order to analyze the climatic variations.

This system can however facilitate the demonstration of the effects of changes quite expressly, as the changes of temperatures and the variations of phytoplankton in sea, the effects of fires and the effective response of the vegetation to drought crisis.

The possibility of integration with measures of spectral signatures on the ground, allows determining the phenological curves of most types of vegetation by satellite images. The availability of spectral signatures facilitates the correction of the atmospheric effects on images, while the measures of specific quantities as humidity, biomass etc. (more than the identification of the physical processes in the atmosphere) can also allow in the short term the continuous surveying of the variations.

5 Conclusions

The formulation of this complete service, composed by satellite data, spectral signatures and a management system of also vector data used through the Internet, opens the doors to a completely different approach to the use of satellite data from a Network of users and laboratories. The approach to the remote data bank also provides the effective possibility to create multiple combinations of images placed side by side, facilitating the interpretation and the comparisons with other products, with the possibility to use one's own data together with geographic vector data made available by data banks.

This new approach makes available, for an immediate use in a large possibility of applications and for a wide users community, the satellite images acquired at Telegis Lab, with the possibility to save locally (or in the server) the vector interpretations.

Our integrated system offers another advantage consisting in the reduction of the times between the acquisition of the data and the access to the information, passing from some days to few hours, by means of centralization of the data acquisition, processing and publication.

The availability via Internet, finally, surely promote and support networking between various subjects and from the complementary interests, increasing the horizon of the possible cues of search and possible applications.

Bibliography

Cheng Y., Gamon J.A., Fuentes D.A., Mao Z., Sims D.A., Qiu H.-l., Claudio H., Huete A., and Rahman A.F. 2006. *A multi-scale analysis of dynamic optical signals in a Southern California chaparral ecosystem: a comparison of field, AVIRIS and MODIS data.* Remote Sensing of Environment 103: 369–378.

Colandrea P. 2004. *Towards new Models of EO Data Exploitation by non expert users.* SSE 2004 Workshop, ESRIN, Frascati (http://earth.esa.int/rtd/ Events/SSE_2004/).

Colandrea P., Amato U., Canfora M.F., Franzese M., Manoni G., Pignatti S., Salvati A., and Serio C. 2004. *WAGRIT: Web per i controlli AGRIcoli e del Territorio*, Proceedings della 8° Conferenza ASITA, Roma 14–17 dicembre 2004.

Dedecker, Ralph G., Whittaker, Tom, Garcia, Ray, K., and Knuteson, Robert O. 2004. *A system design for storing, archiving, and retrieving hyperspectral data*, 20th Conference on IIPS, 84th AMS Annual Meeting, 11–15 January 2004, Seattle, WA. Published by the American Meteorological Society, Boston, MA.

Justice C.O., Vermote E., Townshend J.R.G., Defries R., Roy D.P., Hall D.K., Salomonson V.V., Privette J.L., Riggs G., Strahler A., Lucht W., Myneni R.B., Knyazikhin Y., Running S.W., Nemani R.R., Zhengming Wan Huete A.R., van Leeuwen W., Wolfe R.E., Giglio L., Muller J., Lewis P., and Barnsley M.J., *The Moderate Resolution Imaging Spectro-radiometer (MODIS): landremote sensing for global change research.* Geoscience and Remote Sensing, IEEE Transactions 36(4): 1228–1249.

Marini A., Colandrea P., Melis M.T., Muntoni F., and Sifredi D. 2004. *Firme spettrali di unità territoriali e gestione di immagini MODIS sul Bacino del Mediterraneo*, in Environment Identité Mediterranée – Corte. Julliet 19–25.

Payan W., and Royer A. 2004. *Spectral emissivity of northern land cover types derived with MODIS and ASTER sensors in MWIR and LWIR.* Canadian Journal of Remote Sensing 30(2): 150–156.

GEOSITES AND TOURISTIC DEVELOPMENT OF THE NORTHWESTERN TABULAR MIDDLE ATLAS OF MOROCCO

MOHAMED EL WARTITI[1], AMINA MALAKI[1], MOHAMED ZAHRAOUI[1], FELICE DI GREGORIO[2], AND JO DE WAELE[2]

[1]*Universite Mohammed V, Agdal – Faculte des sciences de Rabat, Département des sciences de la terre, Av-Ibn Batouta, B.P. 1014 – Morocco, wartiti@fsr.ac.ma,*

[2]*Università di Cagliari, Dipartimento Scienze della Terra, Via Trentino, 51 – 09127 Cagliari, Italy, digregof@unica.it*

Abstract

Across the northwestern Tabular Middle Atlas of Morocco there are many examples of landscapes, rocks and fossils providing key evidence of a particular moment or period in Earth history. Such Earth heritage sites are important for educating the general public in environmental matters. They also serve as tools for demonstrating sustainable development and for illustrating methods of site conservation, as well as remembering that rocks, minerals, fossils, soil, landforms form an integral part of the natural world. The significance of certain sites for aesthetic or tourism reasons is obvious. There are numerous geosites, which could contribute to effective exploitation of geotourism, often in conjunction with ecotourism. The strategy employed in such sites involves close consultation with all communities in the vicinity of each geo-site and is not only aimed at tourism and education, but also at sustainable improvement of infrastructures of the people of this area. Geological heritage sites, properly managed, can generate employment and new economic activities, especially in regions in need of new or additional sources of income.

Keywords: Geosites, Tabular Middle Atlas, sustainable development, ecotourism, geotourism

1 Introduction

The area of study (Tabular Middle Atlas) which forms part of the Middle Atlas chain is particularly rich in geosites and geomorphosites of great importance for the study of the geomorphological heritage and geoscientific landscape (Fig. 1).

A geosite is a site or an "area", a few square meters to several square kilometers in size, with geological and scientific significance, whose geological characteristics (mineral, structural, geomorphological, physiographic) meet one or several criteria for classifying it as outstanding (valuable, rare, vulnerable, endangered).

When a special zone includes more than one particularly rare or beautiful and geologically significant feature, it is referred to as a "geopark" (UNESCO, 2000). The features must be representative of the region's geological history, as well as of the events and processes that shaped it (Prichonnet, 2001).

A. Marini and M. Talbi (eds.), *Desertification and Risk Analysis Using High and Medium Resolution Satellite Data*,

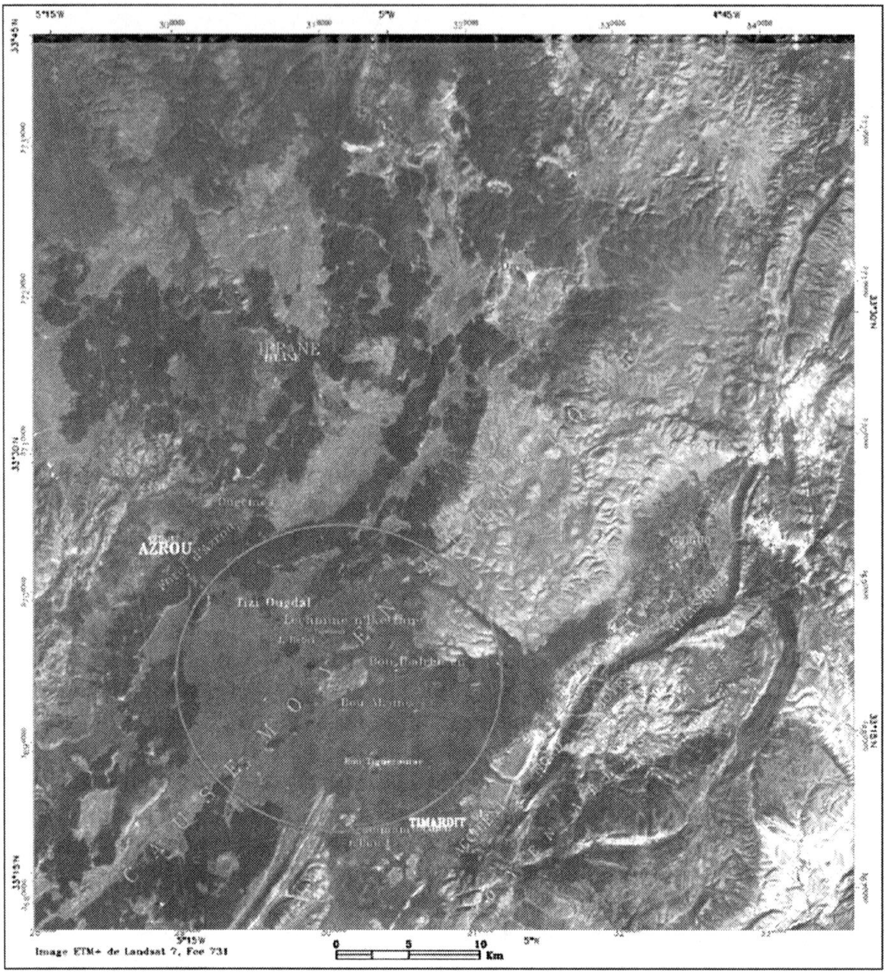

Fig. 1: Space distribution of the geosites inventoried in the Tabular Middle Atlas, on the Landsat/ETM+. Image. UTM projection, WGS-84 ellipsoid

The results of this study represent recommendations of sensitizing geotouristic promotion and protection for this universal natural asset (Di Gregorio et al., 2007).

The Moroccan Atlas system consists of intra-continental mountain ranges. Its evolution was controlled by the opening of the Atlantic and by the Alpine orogeny. Fault systems along strike of these mountain ranges played a major role in this process. The Northern Middle Atlas Fault (= Accident Nord Moyen Atlasique: ANMA – sinistral strike-slip fault – divides the Middle Atlas into two units: the northwestern Tabular Middle Atlas (= Causse Moyen Atlasique) (Fig. 2) and the southeastern Folded Middle Atlas (= Moyen Atlas Plissé).

The Middle Atlas is an intra-continental mountain range belonging to Atlas system (Choubert and Marcais, 1952). It is constituted by faults zone bends whose principal ones are directed NE-SW. This mountain range is divided into two units (Fig. 1): the

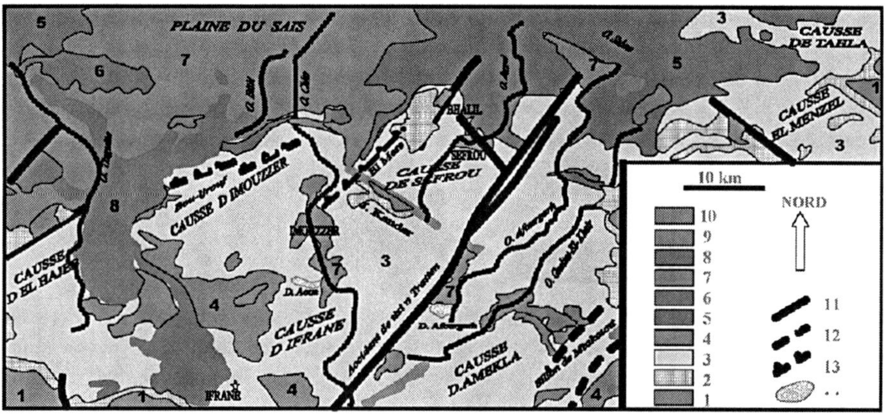

Fig. 2: Structural sketch map of the central part northern Middle Atla: 1: Paleozoic; 2: Triassic; 3: Lower Liassic dolomites; 4: Middle Liassic limestones; 5: Miocene; 6: Pliocene conglomerates and sands; 7: Plio Quaternary; 8: Quaternary basalts; 9: Travertines; 10: Pleisticene fluvial and alluvial fans; 11: Faults; 12: Supposed faults; 13: Flexure; 14: Karstic lakes

northwestern Tabular Middle Atlas (= Causse Moyen Atlasique) and the southeastern Folded Middle Atlas (= Moyen Atlas Plissé), separated by the Northern Middle Atlas fault. Structural shapes indicate two main structures: a large syncline depocenters and straight anticline ridges with local gabbroic intrusions.

The mainly NE/SW trending fault zone bends into N/S direction along its central third. Related to the NE/SW, the main trends are the transversal NW/SE-fault trends. The NE/SW strike-slip faults, as well as their NW/SE trending transversal counterparts, both represent hercynian age basement faults affecting sedimentary cover by multiple reactivation since the Triassic. During the Mesozoic and Tertiary the following important events can be distinguished (Rhrib, 1997):

Synsedimentary active normal faults have been proofed along the ANMA, especially for block rotations along normal faults during the Upper Triassic, for intra-formational unconformities during the Carixian and Domerian, and for the generation of a Toarcian halfgraben (Rhrib, 1997).

Slump structures closely related to NE/SW striking faults provide evidence for block tilting during the Dogger. The compressive Cretaceous-Eocene period started in pre-Barremian times and reached one maximum during the Senonian. The development of facies and thickness distribution resembles intense synsedimentary tectonic activities. Basal angular unconformities and conglomerates of variable thickness indicate continuing tectonic activities (Rhrib, 1997).

Tertiary thrusts developed during renewed compression, which are contemporaneous with decollement of the sedimentary cover of the Qued Zra-Block (Rhrib, 1997).

The paroxysmal character of the compression tectonics during Upper Miocene and Quaternary can be confirmed through the development of mega tension fractures in the Middle Atlas by blocking the NE-SW sinistral strike-slip movements. These fractures reach the deeper lithosphere and cause rising of magma (Rhrib, 1997).

The evolution of longitudinal NW/SE trending Miocene basins was related to extensional fault systems, that controlled the development of halfgrabens, thus locking strike-slip motions along the ANMA (Rhrib, 1997). Such motions are also indicated by the occurrence of asymmetric Z-folds with vertical axes in Upper

Eocene rocks from the Ain Nokra syncline as well as amygdale-structures from Jbel Hayane. Furthermore, the amount of lateral transport (7 km) along the ANMA can be estimated due to the discovery of a tectonic friction-breccia 20 km south of Taza, consisting of Toarcian carbonate and basalt components supported by a matrix of red Triassic pelites. Furthermore, the Z-folds permit the calculation of the main deviatoric paleostress directly at the ANMA (Rhrib, 1997).

2 Geosite of karstic landforms

In the Middle Atlas mountains, one can count more than 40 dayats (lakes) corresponding to tectono-karstic collapses with fluvio-lacustrine deposits. The existence of these plio-quaternary basins in the tabular, allows us to analyze the tectonic episodes responsible of their genesis and their evolution. This study focuses on the structural evolution to contribute to the comprehension of their genesis and their evolution, and to establish a chronology of the tectonic phases dating from the study of syrisedimentary structures and kinematics analysis of microtectonic sites (Angelier and Goguel, 1979; Carey, 1979).

Synthetic lithostratigraphic columns of the plio-quaternary fluvio-lacustrine beds were established dating from the detailed logs that we realized in the deposits of the lakes Aoua and Afourgagh; which shows the best outcrops of the lacustrine basins of the Middle Atlas. Two different groups of deposits are recognizable: the marginal facies and the central facies.

A morphostructure of lakes Aoua, Afourgagh, Ifrah, and Aguelmam Sidi Ali shows us that the morphology of these basins is related to fault intersections, whose major directions are N 135°, N 080°, N 030° and N 170°. The creation and the multiple reactivations of these faults are attribute to Upper Miocene-Quaternary. The effects of these faults are amplified by the karstification of liasic carbonates.

The analysis of the brittle deformation in these basins deposits and in the Jurassic substratum allows us to characterize four principal tectonic episodes responsible for the genesis and evolution of these fluvio-lacustrine basins. The first episode is an Upper Miocene NE-SW extension. The second is a Mid-Upper Pliocene NNW-SSE extension. The two last Quaternary episodes correspond to stress fields with $\sigma 3$ directed N 120° in the Lower-Mid Quaternary, and with $\sigma 3$ directed N 080° in the Mid-Upper Quaternary.

2.1 Aguelmam Afourgagh (Dayet Afougha/Afourgah)

The lake catchments (49 km^2) (Fig. 3), mainly made of Triasic silts and Liasic dolomites (about 63% of the surface), is located in a karstic environment. Karst and the local tectonics induced the formation of closed depressions where groundwater crops out (Chillasse et al., 1999).

The lake Afourgagh (6 ha), located in the Middle-Atlas mountains (Morocco), showed a strong decrease of its water level during the last 20 years (6 m between 1987 and 1996) (Flower et al., 1989), inducing consequences on the life of local population. This recent decrease of the water level allows observing and studying the most recent (Holocene) lake deposits.

Fig. 3: Photo of Aguelmam Afourgagh. Details of Dayat Afourgah visited during the current study, showing latitude (N) and longitude (W), area in May 1999 (ha) and conductivity (mS) in October 1997 and May–June 1999

WETLAND	LATITUDE	LONGITUDE	AREA	Cond 97	Cond 99
Dayat Afourgah	33°37'	04°53'	5	1.6	

The lacustrine records consists of interbedded pluri-decimetric white layers of charophytes tufas containing, sometimes, a lot of well preserved encrustated stems and gyrogonites that show the strong authigenic production, pluri-centimetric dark layers (with samples containing a lot of organic components) and silty-clayey detrital layers. The tufas are mainly made of calcite, high magnesium calcite and aragonite that can account for up to 81% of the bulk. Aragonite shows that the water was probably rich in sulfate and magnesium.

The sediments present variable amounts of gyrogonites and oogonium of characeae, ostracods and molluscs. Studying these organisms gives a lot of clues about the recent evolution of the lake: water level fluctuations, water chemistry, hydrology, climate.

The study of the geometry of the deposits coupled to the study of the different facies make possible to reconstruct the lake evolution during the Holocene and to discern the natural and anthropic factors responsible of the recent hydrological variations.

During the 1970s and early 1980s, this was the most important of the Middle Atlas lakes for breeding and wintering water birds (Thévenot, 1976), and was still a clear-water lake in good condition (with a conductivity of 0.75 mS) in 1984 (Flower, 1989), yet by 1987 all the reed beds had been removed by cutting and grazing, many water bird species had disappeared and numbers of wintering water birds had crashed (Franchimont, 1994). It is now greatly reduced in area, highly turbid, totally devoid of submerged vegetation and with very few water birds (only 31 cattle egrets and 4 other birds on 9.10.97). Half of the shoreline is now surrounded by arable land. In 1997 this lake had an exceedingly low water level, and only c.30% of the basin was flooded. Exotic fish include *Esox lucius* and black bass *Micropterus salmoides* (Chergui, 1999). Marbled teal were formerly present at this site (with 150 in September 1969, Thévenot in litt., 1976) but have not been recorded since 1981 (Green, 1993). Crested coot used to breed there, and up to 300 ruddy shelduck (in November 1978) was recorded in winter (Thévenot, 1976).

2.2 Dayet Aaoua

The Dayet 'Awa is located in a karstic environment (Malaki, 2006) (Fig. 4). Karst and the local tectonics induced the formation of closed depressions where groundwater crops out (Chillasse et al., 1999). The Dayet 'Awa (140 ha), located in the Middle-Atlas mountains (Fig. 3) (Morocco), remains a relatively well-conserved and diverse lake of great conservation importance. The vegetation community remains rich, although there is no fringe of *Phragmites* and *Scirpus lacustris* along the southwest shore as described by Morgan, probably due to overgrazing and reed cutting (Franchimont, 1994). However, these and other emergent plants are found at the eastern end. Pumping of subterranean water from the underlying aquifer leads to low water levels in years of dry rainfall (Dakki and El Hamzaoui, 1997). "Pedalo" boats are no longer in use, although local tourism (picnickers from Fes and Meknes) has intensified (Franchimont, 1994). The development of intensive chicken farms in the catchments may lead to significant pollution. This lake is enormously important

Fig. 4: Photo of Dayet Aaoua. Details of Dayat Aoua visited during the current study, showing latitude (N) and longitude (W), area in May 1999 (ha) and conductivity (mS) in October 1997 and May–June 1999

WETLAND	LATITUDE	LONGITUDE	AREA	Cond 97	Cond 99
Dayat Aoua	33°39'20"	05°02'	50	0.5	0.42

for crested coot, and 1,200 marbled teal were recorded in April 1999. Morgan underestimated the area of this lake, which is 140 ha (El Agbani, 1997). Exotic fish include *Esox lucius* and *Tinca tinca* (Chergui, 1999).

The latest news from September 2001 is that this lake was completely dry. It is unclear whether or not this is a "natural" event due to reduced precipitation, or to what extent increased water extraction for agriculture may be responsible. It is also unclear whether the lake will flood again in the near future or not.

2.3 Dayet Ifrah

This lake (Fig. 5) has been highly degraded since Morgan's time. Rapid sedimentation from the surrounding steep slopes now denuded of trees is the probable cause of a reduction in the lake area and high turbidity. As a result, submerged vegetation has been eliminated except for tiny patches of *Ranunculus*. Numbers of wintering water birds have decreased steadily since 1983 (Franchimont et al., 1994). This site has lost its former importance for marbled teal, of which 300 were counted in 1973 (Green, 1993). Exotic fish include *Esox lucius*, *Rutilus rutilus*, *Phoxinus phoxinus* and *Perca fluviatilis* (Chergui et al., 1999).

Fig. 5: Photos of Dayet Ifrah. Details of Dayat Ifrah visited during the current study, showing latitude (N) and longitude (W), area in May 1999 (ha) and conductivity (mS) in October 1997 and May–June 1999

WETLAND	LATITUDE	LONGITUDE	AREA	Cond 97	Cond 99
Dayat Ifrah	33°33'45"	04°55'45"	100	0.74	0.96

2.4 Aguelmam Sidi Ali

Considered by Morgan (1982) as "fresh", we recorded conductivities of 1.1–1.6 mS, equivalent to oligosaline. Water levels are now reduced (Fig. 6) and the marsh to the southwest of the lake described by Dorst (1951) no longer exists. Numbers of wintering water birds have decreased markedly since 1983, and poaching of the protected *Tadorna ferruginea* and other waterfowl is a major problem (Franchimont et al., 1994).

Grazing pressure in spring and summer by flocks of sheep and goats brought by nomads is intense, and emergent vegetation is almost eliminated. The nesting of *Podiceps cristatus* as described in 1965 (Géroudet, 1965) occurred up to at least 1985 (Thévenot, 1976) but is now impossible owing to lack of vegetation. The native grayling-trout *Salmo pallaryi* became extinct after the introduction of the carp *Cyprinus carpio* in 1934 (Chergui, 1999). Other exotic fish include *Esox lucius*, and *Stizostedion lucioperca* (Chergui, 1999).

Fig. 6: Photo of Aguelmame Sidi Ali. Details of Aguelmam Sidi Ali visited during the current study, showing latitude (N) and longitude (W), area in May 1999 (ha) and conductivity (mS) in October 1997 and May–June 1999

WETLAND	LATITUDE	LONGITUDE	AREA	Cond 97	Cond 99
Aguelmam Sidi Ali	33°04'34"	05°00'13"	150	1.4	1.1-1.6

2.5 Drought climates changes and desertification risk in the tabular Middle Atlas wetlands

The tabular Middle Atlas is the most important water reservoir in Morocco. Of all the North-African mountains, it is the richest in wetlands, especially natural lakes, cold rivers and springs. These aquatic ecosystems fulfill significant hydrological, socio-economic and ecological functions, while their biodiversity classifies them as internationally important wetlands, as for the Ramsar Convention. However, most of these wetlands are suffering from several anthropic disturbances, which are amplified by repetitive natural droughts (Chillasse and Dakki, 2004).

Across the tabular Middle Atlas region, natural wetlands are being destroyed and degraded at a rapid rate, with severe consequences for the status of aquatic fauna and flora (Pearce and Crivelli, 1994). There is an urgent need for more studies of wetland loss, as well as prioritization of remaining wetlands for conservation actions (Jones and Hughes, 1993).

It is conceivable that climate change may have some role in the current loss of Moroccan wetlands. Franchimont et al. (1994) considered that a prolonged drought period since the early 1980s is partly responsible for the disappearance of lakes in the Middle Atlas region. Variation in precipitation clearly influences water levels in these lakes (Flower and Foster, 1992).

Indeed, the recurrent drought, resulting from climate change and variability, accelerates desertification, which in turn contributes to the persistence of drought. Various scientists, scientific institutions and implementing agencies, have suggested a numbers of definitions of desertification. Based on special studies and extensive discussions at the UNEP Consultative Meeting on the Assessment of Desertification in Nairobi in 1990, the following new definition was suggested: "Desertification is Land Degradation in Arid, Semi-arid and Dry Sub-humid Areas resulting from various factors, including climate variation and human activities. Land in this concept includes soil and local water resources, land surface and vegetation or crops. Degradation implies reduction of resource potential by one or a combination of processes acting on the land" (UNEP, 1996). This definition implied agreement on a tool for assessing and combating desertification that would be more suitable operationally.

2.6 Ruiniform landscape

Tidirine is a wide and astounding area of ruiniform rocks (Malaki, 2006) (Fig. 7), born out of the limestone plateau dating back to 200 to 140 millions years. Its landscape's feature is a series of ruiniform rocks. Winds and rains had made their work on those rocks, by eroding them (Elbouqdaoui et al., 2007), providing them with strange shapes (Malaki et al., 2007a, b).

Fig. 7: Photos of Tidirine (Malaki, 2006)

3 Geosite of volcanic landforms

The basaltic plateau of Azrou is characterized by its unique volcanic landscape (Malaki et al., 2007c, d) (Fig. 8), as well as botanical and faunal diversity. This Quaternary-Age basaltic volcanic complex is also one of the most diverse in North Africa, including Aa and, Cinder Cones, composite or stratovolcano, caldera, crater lake and plateau basalt (Malaki, 2006).

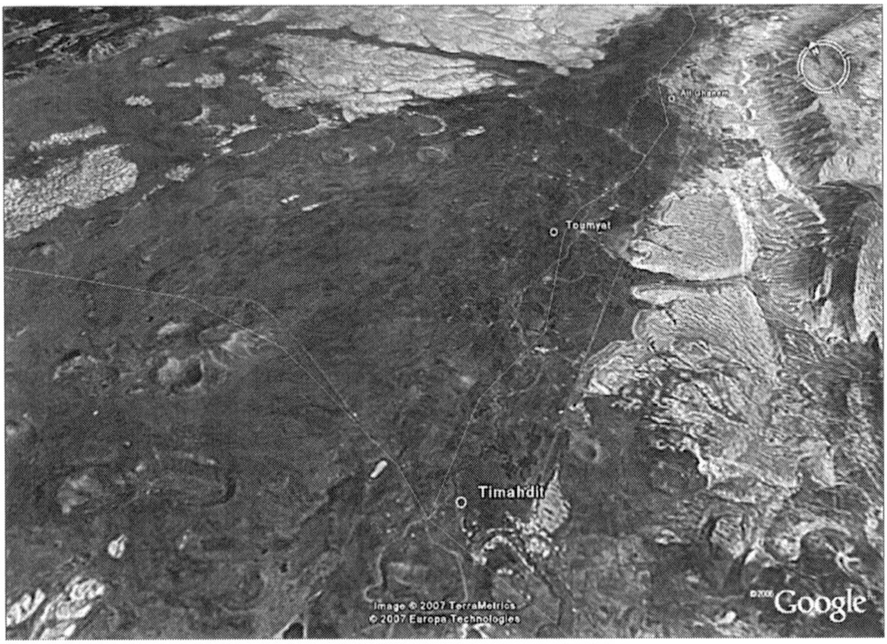

Fig. 8: The volcanic landscape of the plateau of Azrou. Satellite image provided by Google Maps service

3.1 Bou Tegrrouine

Bou Teguerrouine (Fig. 9) is the result of alternating eruptions of pyroclastic material and lava flows (Fig. 10). It results in alternating layers of lava and volcanic ash. The lava flows protect underlying ash deposits from erosion (Kharbouch et al., 2007).

Due to their relatively high viscosity, the flows cannot flow great distances from their source. Consequently composite volcanoes are typically high and have very steep slopes. Due to high magma viscosity composite volcanoes are characterized by highly explosive eruptions, producing a greater volume of pyroclastic material than lava flows. This composite volcano has a crater at the summit, which contains a clustered group of vents. Lavas either flow through breaks in the crater wall or issue from fissures on the flanks of the cone. Lava, solidified within the fissures, forms dikes that act as ribs which greatly strengthen the cone.

Fig. 9: Photo of Bou Teguerrouine

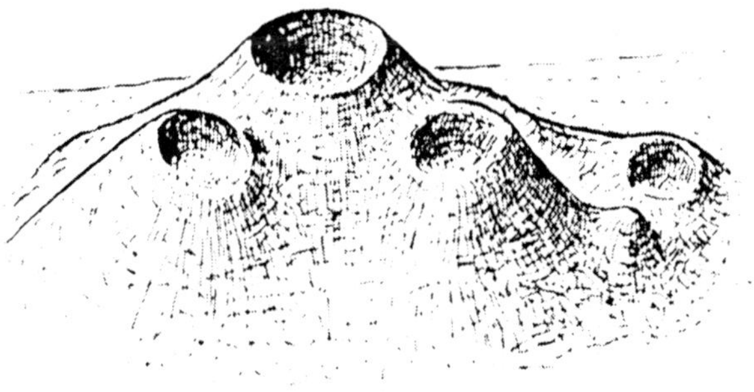

Fig. 10: Schematic representation of Bou Teguerrouine

3.2 The caldera of the Michliffen

The caldera of the Michliffen is a circular depression (Figs. 11 and 12) at the summit of a volcano. It was formed when magma was withdrawn or erupted from a shallow underground magma reservoir. The removal of large volumes of magma may result in loss of structural support for the overlying rock, thereby leading to collapse of the ground and formation of a large depression.

The largest and most explosive volcanic eruptions eject tends to hundreds of cubic kilometers of magma onto the Earth's surface. When such a large volume of magma is removed from beneath a volcano, the ground subsides or collapses into the emptied space, to form a huge depression called a caldera.

4 Conclusion

Many geosites of the northwestern Tabular Middle Atlas are not included in protected natural areas, often due to a lack of knowledge of the existence of these values.

As a prior step to the scoring of measures to preserve geosite, a database must be created that is accessible and comprehensible to all organizations involved in one way or another in territorial management. The main objective is to make the

Fig. 11: Photos of the Michliffen

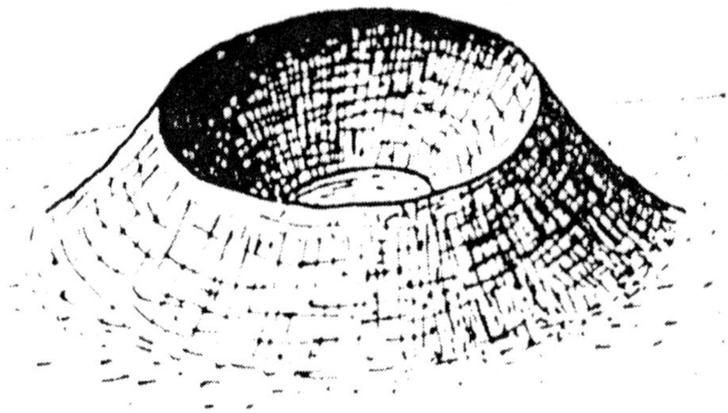

Fig. 12: Schematic representation of Michliffen

necessary information on geological heritage available, both regarding its location and the values that characterize it, as well as indicating which actions may put them at risk and which may favor their conservation. This information is directed at organizations involved in the planning and decision-making processes that may affect on natural heritage. The database must be integrated in the Natural Heritage Information System.

In order to attain this objective, an inventory has been created containing all those outcrops, isolated or in groups, in areas of variable size, which are considered basic and representative. The group of areas of geological interest constitutes a record of the geological evolution of the northwestern Tabular Middle Atlas territory.

The information included in the Inventory of geosites of the northwestern Tabular Middle Atlas is available to the general public for its use in different circumstances, always in agreement with the main objective of preserving geological heritage:

- Protection of an area within the framework of a particular action. In this way, any actions on the physical environment can be planned at any level, from the private or government sector, ensuring that such actions respect geological heritage and negative effects are avoided.
- Promotion of the area in order to attract visitors. Moreover, the "exploitation" of this resource would have positive repercussions on the dissemination of geological values and the need for geoconservation. Many organizations would benefit from the acquisition of knowledge of geological heritage and its location. We propose the creation of several tourist circuits, which are the basis for a further valorization of this territory.

Bibliography

Angelier J. and Goguel J. 1979. *Sur une méthode simple de détermination des axes principaux des contraintes pour une population de failles.* Comptes Rendus de l'Académie des Sciences, Paris, 288, n.**3**, Sér. D, 307–310.

Carey E. 1979. *Recherche des directions principales de contraintes associées au jeu d'une population de failles.* Revue de Géographie Physique et Géologie Dynamique, 21, n.**1**, 57–66.

Chergui H., Pattee E., Essafi K., and Mhamdi M.A. 1999. *Moroccan limnology*, in Wetzel, R.G., Gopal, B. (Eds.), Limnology in Developing Countries, n.**2**. International Scientific Publications, New Delhi, pp. 235–330.

Chillasse L. and Dakki M. 2004. *Potentials and conservation status of the Middle Atlas wetlands (Morocco), and the impact of the drought.* Sécheresse, 15, n.**4**, 337–345.

Chillasse L., Dakki M., and Abbassi M. 1999. *Les lacs naturels du Moyen Atlas, zone d'intérêt majeur pour la conservation de la biodiversité*, in Proceeding of First International Conference on Biodiversity and Natural Resources Preservation, School of Science & Engineering, Al Akhawayn University, Ifrane, May, pp. 13–14, 19.

Choubert G. and Marçais J. 1952. *Géologie de Maroc I. 1re Partie: Aperçu structral, 2e partie: Histoire géologique du domaine de l'Anti-Atlas. Prot. Rép. Franç. Maroc, Serv.* Mines Carte Géol. Notes Mém. 100, 195 pp.

Dakki M. and El Hamzaoui M. 1997. *Rapport National sur les Zones Humides (Maroc).* MedWet report. Administration des Eaux et Forêts et de la Conservatión des Sols, Rabat.

Di Gregorio F., De Waele J., Melis M.T., Fadli D., El Wartiti, and Zarhaoui M. 2007. *Innovative experiences of cultural and environmental goods mapping along the Kasba road on the the Saharan side of hight Atlas.*

Dorst J. 1951. *Observations ornithologique dans le Moyen-Atlas*, in L'Oiseau et la R.F.O., n.**21**, 289–303.

Driss Fadli, Mohamed Bouabdelli, Yves Cailleux, Ahmed El Hassani, Mohamed El Wartiti, Christian Hoepffner, Fatima Kharbouch, Alain Pique, Abdelfattah Tahiri, and Mohamed Zahraoui. 2007. *Geological map of the Moroccan Central Massif; a structural interpretation of the Hercynian Belt.* p.42, Wiget1.2007, Ifrane, Maroc.

El Agbani M.A. 1997. *L'Hivernage des Anatidés au Maroc: principales espèces, zones humides d'importance majeure et propositions de mesures de protection.* Thèse de dctorat d'Etat ès-Sciences, Faculté des Sciences, Rabat, 186 pp.

Elbouqdaoui K., Zahraoui M., Ezzine H., Loukili H., Rouchdi M., El Wartiti M., and Fadli D. 2007. *Evaluation de la sensibilistaion des terres à l'érosion à l'aide de la télédétection et du SIG: cas du basin versant du Sebou (Moyen Atlas,Maroc).* P.41, Wiget1.2007, Ifrane.

Fatima Kharbouch, Mohamed Zahraoui, Mohamed El Wartiti, M'hamed EL Azzouzi, Benacer El Mahi, and Taoufiq M'barki. 2007. *Le Géo-parc volcanique d'Ifrane et d'Azrou: un musée naturel*, p.45, Wiget1.2007, Ifrane, Maroc.

Felice Di Gregorio, Jo De Waele, Maria Teresa Melis, Mohamed El Wartiti, Mohamed Zahraoui, and Driss Fadli. 2007a. *Landscape units, Geomorphosites and Geodiversity of the Ifrane Azrou region (Middle Atlas, Morocco)*, pp. 7–10, Wiget1.2007, Ifrane, Maroc.

Felice Di Gregorio, Mohamed El Wartiti, Mohamed Zahraoui, Jo de Waele, and Amina Malaki. 2007b. *Les Géosites des paysages panoramiques: étude de cas: Panorama d'Itto*. pp. 50–57, Wigetl.2007, Ifrane, Maroc.

Flower R. and Foster I.D.L. 1992. *Climatic implications of recent changes in lake level at Lac Azigza (Morocco)*. Bulletin de la Société Géologique de France 163, n.**1**, 91–96.

Flower R.J., Stevenson A.C., Dearing J.A., Foster I.D.L., Airey A., Rippey B., Wilson J.P.F., and Appleby P.G. 1989. *Catchments disturbance inferred from paleolimnological studies of three contrasted sub-humid environments in Morocco.* Journal of Paleolimnology, 1, 293–322.

Franchimont J., Chahlaoui A., Samih M., and Sayad S. 1994. *Analyse de l'evolution des effectifs des oiseaux d'eau hivernants dans le Maroc Central au cours de la decennie 1983–1993.* Porphyrio, 6, n.**1**, 7–94.

Géroudet P. 1965. Les rapaces diurnes et nocturnes d'Europe. Delachaux et Niestlé. Neuchâtel-Paris, 369–377.

Green A.J. 1993. *The status and conservation of the Marbled teal Marmaronetta angustirostris*. IWRB Special Publication, n. **23**.

Jones T.A. and Hughes J.M.R. 1993. *Wetland inventories and wetland loss studies—a European perspective*, in Moser, M, Prentice, R.C., van Vessem, J. (Eds.), Waterfowl and Wetland Conservation in the 1990s—A Global Perspective. Proc. IWRB Symp., St. Petersburg Beach, Florida. IWRB Special Publication No. 26, Slimbridge, pp. 164–169.

Malaki A. 2006. *Géosites: Intérêt scientifique, patrimoine culturel et visées socio-économiques, au niveau d'Ifrane, Azrou, Aîn leuh et El Hajeb (causse moyen atlasique).* Thèse de doctorat national en géologie, Faculté des Sciences de Rabat, 278 pages.

Malaki A., Di Gregorio F., El Wartiti M., De Waele J., and Zahraoui M. 2007a. *Cas d'un Géosite hydrique à protéger: Cascades d'Ifrane-Zaouia.* pp. 58–64, Wigetl.2007, Ifrane, Maroc.

Malaki A., El Wartiti M., Di Gregorio F., Zahraoui M., Kharbouch F., El Mahi B., and Fadli D. 2007b. *Pour une conservation et une valorisation des géomorphosites volcaniques du causse Moyen Atlasique*. pp. 75–90, Wigetl.2007, Ifrane, Maroc.

Malaki A., El Wartiti M., Di Gregorio F., Zahraoui M., and De Waele J. 2007c. *Les Geosites Des Paysages Panoramiques: Etude De Cas: Panorama D'itto.* pp. 50–57, Wigetl.2007, Ifrane, Maroc.

Malaki A., El Wartiti M., Di Gregorio F., Zahraoui M., Kharbouch F., El Mahi B., and Fadli, D. 2007d. *Pour Une Conservation Et Une Valorisation Des Geomorphosites Volcaniques Du Causse Moyen Atlasique*. pp.75–90, Wigetl.2007, Ifrane, Maroc.

Morgan N.C. 1982. *An ecological survey of standing waters in North West Africa: III. Site descriptions for Morocco*. Biological Conservation, n.**24**, 161–182.

Pearce F. and Crivelli A.J. 1994. *Characteristics of Mediterranean Wetlands*. n MedWet publications, n**1**. Tour du Valat.

Prichonnet G. 2001. *La notion de géosites et de géoparcs et son application à la Stratégie québécoise sur les aires protégées*, in Association professionnelle des géologues et géophysiciens du Québec, 12 pages.

Québec, Ministère De L'environnement (MENV). 2002. *Les aires protégées au Québec: une garantie pour l'avenir*, Plan d'action stratégique, premiers résultats, 44 pages.

Rhrib J. 1997. *Die Störungszonen des Mittleren Atlas (Zentralmarokko): strukturelle Entwicklung in einem intrakontinentalen Gebirge = Les zones des failles dans le Moyen Atlas (Maroc), leurs évolutions structurales dans une chaine intracontinentale*/Jamal Rhrib. – Berlin: Selbstverl. Fachber. Geowissenschaften, FU Berlin, 1997. - 221 S.: Ill., graph. Darst. (Berliner geowissenschaftliche Abhandlungen: Reihe A; 194.)

Thévenot M. 1976. *Les oiseaux de la réserve de Sidi Bou-Rhaba*, Bull. Inst. Sci. Rabat, 1, pp. 67–99.

UNEP. 1996. *Report of the Regional Consultations held for UNEP's first Global Environment Outlook.* Nairobi.

UNESCO. 2000. *Report by the Director-General on the feasibility study on developing a UNESCO Geosites/Geoparks Programme*, Executive Board, 160th session.

MONITORING LAND DEGRADATION IN DRYLANDS BY REMOTE SENSING

WEICHENG WU
International Center for Agricultural Research in Dry Areas (ICARDA), Aleppo, Syria
w.wu@cgiar.org

Abstract

This article aims at introducing the application of remote sensing techniques to monitoring land degradation and desertification in arid regions based on a review on actually available methods and several pertinent case studies. It is considered that land degradation, a process of reduction in vegetation cover and water resource, soil erosion and of salinization, etc., is a subtle and progressive environmental change in time. It is therefore necessary to conduct multi-temporal or even time-series observation. Coarse resolution data can be used to reveal regional, continental and even global level environmental changes and target the "hotspots". However it leaves many uncertainties. On the contrary, high and very high spatial resolution data are capable for highlighting such a subtle change in detail at local level. To avoid bad-interpretation, meteorological data should be combined in analyses to make sure that the differences in spectral reflectance observed are representing true changes but not climate related events like droughts. Furthermore, it is essential to link remote sensing with human activity to understand the mechanism of land degradation and its driving forces. In this way, remote sensing will be not only a powerful tool for providing dynamic information to monitor land surface changes and degradation on different scales but also for helping decision-making in producing relevant mitigation measures for sustainable resource exploitation.

Keywords: Activedunes, Desertification, LandDegradation, VegetationIndex, Vulnerability

1 Introduction

As Thomas and Middleton (1994) said, monitoring is one of the important procedures in the arid land research. In fact monitoring land degradation and desertification is not a new subject. Since Lamprey (1975) reported his findings in UNESCO/UNEP that the Sahara Desert in North Sudan had encroached southward at a rate of 5–6 km/year, desertification and land degradation monitoring and assessment have become a hotspot in dryland research. Since then numerous international, governmental and non-governmental institutions and individual scientists have heavily invested in this investigation. With the contribution of hundreds scientists, especially

A. Marini and M. Talbi (eds.), *Desertification and Risk Analysis Using High and Medium Resolution Satellite Data*,

that of Hellden (1988 and 1991) and Tucker et al. (1986 and 1991) based upon remote sensing observation, the myth of "Sahara southward encroachment" was broken. In the report of Tucker et al. (1991), the extension and contraction of the Sahara Desert or the forth-and-back movement of the southern Sahara margin is a climate related phenomenon. When precipitation increases, the surface of the Desert contracts and when it decreases, the Sahara extends. Thus a world level controversy about the desertification and desert encroachment was put a point. There was no desert progression or "marching desert" (Forse, 1989) as Lamprey (1975) reported. However, this does not mean that land degradation, a phenomenon of continuous process in reduction of vegetation cover and vigor, water resources, soil erosion and salinization, does not exist (Smith and Koala, 1999; Wu 2003a). The importance lies in how to discern and reveal such a subtle change.

Remote sensing, thanks to its advantages in providing dynamical, multi-temporal or time series land cover information, has been widely applied in dryland research including discriminating land use changes and land degradation since the 1970s (Courel et al., 1984; Tucker et al., 1986, 1991; Graetz et al., 1988; Hellden, 1988, 1991; Hanan et al., 1991; Lambin and Strahler, 1994; Gao et al., 2001; Wu, 2003a–c and 2004; Wu et al., 2002 and 2005).

However measuring dryland degradation is particularly difficult because there is strong interaction between erratic and natural in rainfall and anthropogenic changes in vegetation cover (Lambin, 1997). Therefore, monitoring is an intensive research in which any factor leading to misunderstanding and confusion, e.g., vegetation cover change due to temporal climate fluctuation must be taken into account. In other words, monitoring and assessing land degradation by remote sensing should be integrated with an analysis on the meteorological data to make sure that the changes observed in spectral reflectance and photosynthesis are not just a temporal phenomenon caused by climate event like drought but a true progressive degradation in vegetal greenness and biomass. Thus monitoring cannot be focused only on a bi-temporal space but on a multi-temporal or even time-series space.

Based on a review of the actually available algorithms and procedures, this paper is to unfurl and generalize the remote sensing application to land degradation by several case studies of the author.

2 Monitoring algorithms and procedures

Generally the approaches of remote sensing application to land degradation monitoring can be depicted as follows:

2.1 Data selection

An important phenomenon in remote sensing is the scale effect. The detail extent of land surface features discerned is different while we observe them on different scales or by "zooming" (Raffy, 1994). Each scale of measurement requires correspondingly its optimal pixel resolution. This is the so-called scale effect in remote sensing (Woodcock and Strahler, 1987; Quattrochi et al., 1997; Wu, 2007). So for different scales of research we should choose different spatial resolution data (Table 1).

Table 1: Frequently used remote sensing data and their suitability for study

Sensor	Spatial resolution	Band number	Suitability	Available since
QuickBird	0.6–3 m	5 bands	Local study	2001
IKONOS	0.8–4 m	5 bands	Local study	1999
SPOT HRV and HRG	2.5–20 m	3-5 bands	Local study	HRV 1986 and HRG 2002
IRS	2.5–72.5	2-4 bands	Local and regional	1988
Landsat	15–60 m	7-8 bands	Local and regional study	MSS 1972 and TM 1984
ASTER	15–90 m	14 bands	Local and regional study	1999
CBERS	20–260 m	11 bands	Local and regional study	1999
MODIS	250–1 km	36 bands	Regional and continental	1999
SPOT-Vegetation	1 km	5 bands	Regional, continental and global	1998
SeaWiFs	1.1 km	8 bands	Continental and global	1997
AVHRR	1.1–4.1 km	5 bands	Continental and global	1981

Coarse resolution data can be used for large-scale regional and continental monitoring studies. In fact, Justice (1986), Malingreau et al. (1989), Tucker et al. (1991), Hill (1993), Skole and Tucker (1993), Lambin and Strahler (1994), Hill et al. (1995), Lambin and Ehrlich (1997), Lupo et al. (2001) and Young and Wang (2001) have investigated land surface changes in Africa, Mediterranean, South America and China with coarse resolution data (e.g., AVHRR, SPOT Vegetation). Such data can reveal regional scale change and target hotspots but leave many uncertainties, which need further verification.

As for local level studies, a prime requirement on the remote sensing data lies in their capacity to highlight as exactly as possible the phenomenon as land use change and degradation (Wu, 2007). They need high or even very high-resolution data. Also, a great number of authors as Courel et al. (1984), Hellden (1988 and 1991), Graetz et al. (1988), Mertens and Lambin (2000), Gao et al. (2001), Li et al. (2003), Wu (2003a–c and 2004) and Wu et al. (2002, 2005) have carried out such kind of land degradation detection using high spatial resolution data (e.g., Landsat and SPOT). The advantage of high-resolution data lies in the fact that they allow us to understand what has taken place and what is in progress.

Besides the scale effect, another factor to be considered is the availability of data, especially, the ancient ones. Landsat family has been set into service since 1970s and SPOT since 1980s. For decade-level local study, these kinds of data would first come into choice due to their large observation time-span and high spatial resolution.

Sources:

http://www.digitalglobe.com
http://www.spaceimaging.com
http://www.satimagingcorp.com/
http://www.spot-vegetation.com
http://www.spotimage.fr
http://landsat7.usgs.gov
http://www.gds.aster.ersdac.or.jp/
http://www.fas.org/spp/guide/india/earth/irs.htm
http://delenn.gsfc.nasa.gov/~imswww/pub/imswelcome/

2.2 Pre-processing and multi-spectral transformation

It is common knowledge that images acquired by satellites are influenced by atmosphere. It is necessary to carry out first an atmospheric correction by either atmosphere simulation and *in-situ* measurement approaches or image-based method (Dave, 1980; Price, 1987; Kaufman and Sendra, 1988; Kaufman and Tanré, 1992 and 1996; Chavez, 1996; Wu, 2003a and 2004). Merely *in-situ* measurement data and correction programme are not always available, so image-based methods like COST model (Chavez, 1996; Wu, 2003a and 2004) are more applicable. We can also simply conduct a relative correction (Caselles and Garcia, 1989; Wu, 2007).

In addition, during the movement of satellite and topographic effect, some deformation will be produced in images. Thus they need a geometric correction.

Up to date, about 150 multi-spectral transformations have been published among which tens are of scientific value and frequently applied. According to author's experience and opinion, the pertinent ones for dryland study are listed below:

Tasseled Cap Transformation (Kauth and Thomas, 1976; Crist and Cicone, 1984a, b) can convert land cover information included in multispectral bands of Landsat images into three thematic indicators: Brightness, Greenness and Wetness (for MSS Yellowness), which can respectively reveal soil brightness, vegetation vigor and soil moisture. The fourth component from this transformation on TM image is also useful since it can be used to estimate atmospheric effect – haze (Crist and Cicone, 1984b; Wu, 2003a and 2004).

SAVI and EVI: SAVI (Soil-Adjusted Vegetation Index) was proposed by Huete in 1988 based on measurements of cotton and range grass canopies with dark and light soil backgrounds by adding an adjustment factor L into NDVI. For low vegetation cover in the dryland, L can be set to 1. In 1997, Huete et al. improved SAVI and introduced the Enhanced Vegetation Index (EVI) by employing the blue band to correct the atmospheric affectation on the red band in SAVI. This new index EVI is therefore resistant to both soil influences and atmospheric effects and valuable for arid-land research.

ARVI (Atmospherically Resistant Vegetation Index) was developed by Kaufman and Sendra (1988) and Kaufman and Tanré (1992 and 1996). A self-correction process automatically corrects the atmospheric effect in the red band: introducing the blue band into NDVI. It is applicable for both TM and MODIS data.

NDVI (Rouse et al., 1974; Tucker, 1979), although affected by soil reflection and atmospheric effect, it is the most widely accepted and applied index in remote sensing multi-spectral transformation. An advantage of NDVI lies in its simplicity and sensitivity to vegetation cover change.

Spectral Mixture Analysis (SMA) developed by Smith et al. (1990) and Hill (1993) can reveal land surface characters like soil, rock and vegetation by spectral unmixing processing through end-member selection. This transformation can be applied for land degradation assessment by measuring changes in soil/rock ratio and vegetation abundance in time.

Principal Components Analysis (PCA), through a linear transformation on multi-band remote sensing images, can produce a series of orthogonal components and reduce the volume without much information loss (Byrne et al., 1980; Lillesand et al., 1994; Richards and Jia, 1999).

2.3 Selection of suitable algorithm for change detection

Actually, several approaches are available for extracting land use change and land degradation information. They are shown as follows:

Post-classification comparison (Swain, 1976; Gordon, 1980; Singh, 1989; Skole and Tucker, 1993; Lillesand et al., 1994). This is probably the most traditional approach for change detection based on different independent classifications on the multi-data images followed by a comparison between the same classes of different data to identify the changes in land use and land cover including degradation. As many authors argued, this algorithm may produce huge error if each class is not well classified and validated. However, with the improvement of classifiers (such as Maximum Likelihood, Neural Network, etc.) and advancing of computing technology, more and more successful studies have been achieved with this approach (Tucker et al., 1991; Skole and Tucker, 1993; Mas, 1999).

Differencing, Rationing, Regression and Changing vector analysis (CVA) (Engvall et al., 1977; Malila, 1980; Ingram et al., 1981; Jensen and Toll, 1982; Singh, 1989; Lambin and Strahler, 1994). These algorithms can be applied directly on image band or transformed indicators like EVI, ARVI, NDVI, TC features, SMA end-member, and so on. These algorithms involve a threshold technique and thus can localize exactly the positive and negative changes and land degradation highlighted by spectral or vegetation indicator difference. However, an interpretation on the concrete change types is needed based on field investigation or first-hand knowledge.

Cross-correlation and Cross-tabulation analysis (Koeln and Bissonnette, 2000; Hurd et al., 2001). These are two recently developed methods based on classification and threshold. It has been said that they can produce accurate change detection.

Post-classification differencing. This algorithm is proposed based on the author's 10 years practice in change detection. It is in fact a combination of classification and differencing. More concretely, it is first to individually classify the images of different data and then apply a differencing on the same class between different data. In this way, an increase or a decrease of the observed class can be easily underlined.

Advantage: there is no need to set thresholds to extract changes and this algorithm is able to locate precisely the changes in space avoiding the shortcoming of the Post-classification comparison approach.

Requirement: similarly to the Post-classification comparison method, the classification should be done as accurate as possible.

The following section demonstrates three case studies undertaken in the arid regions in China based in a combination of the above algorithms. The location of these sites is shown in Fig. 1.

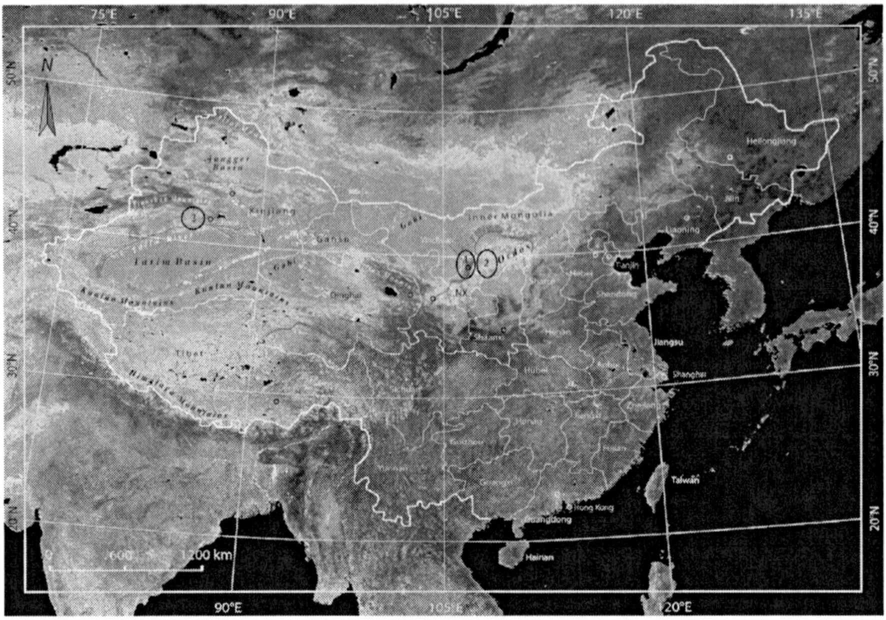

Fig. 1: Location of the case study sites in China (Modified from Wu, 2003a) (1) The region of Yinchuan, (2) West Ordos, and (3) Middle Tarim

3 Examples

3.1 Case study 1

The region of Yinchuan is a dry area where annual rainfall fluctuates between 82 and 262 mm. Agriculture has been well developed due to the irrigation system from the Yellow River. Grazing has occurred mainly in sandy grassland and man-managed pastures. Coal mining in the Helan Mountains and coal-related and chemical industries have been rapidly developing in the past 50 years, especially, since 1980s. Population grows with economic development.

Data used:

1. Multitemporal Landsat images as follows:

129-33-1987 Sept 20, TM; Azimuth: 140.83, Sun elevation: 46.03
129-33-1989 Sept 17, TM; Azimuth: 141.83, Sun elevation: 47.51
129-33-1991 Aug30, TM; Azimuth: 130, Sun elevation: 50
129-33-1999 Aug12, ETM+; Azimuth: 132.73, Sun elevation: 58.60

2. Socio-economic data and meteorological data at county-level

Method and procedures involved:

- Geometric correction based on topographic maps (1/200,000–250,000) and GPS points
- Atmospheric correction by COST model (Chavez, 1996; Wu, 2003a and 2004)
- Tasseled Cap Transformation (Crist and Cicone, 1984a, b) and derivation of the indicators: Brightness and Greenness

- Brightness and Greenness differencing and threshold to produce 3-Class change maps: positive change, negative change and no-change
- Change interpretation and concrete change type identification based on field investigation

3.1.1 Mapping changes and quantification: Results

The results revealed by the above processing are shown in Fig. 2. Totally 11.7% of this dry area has been modified due to different kinds of land use. The most

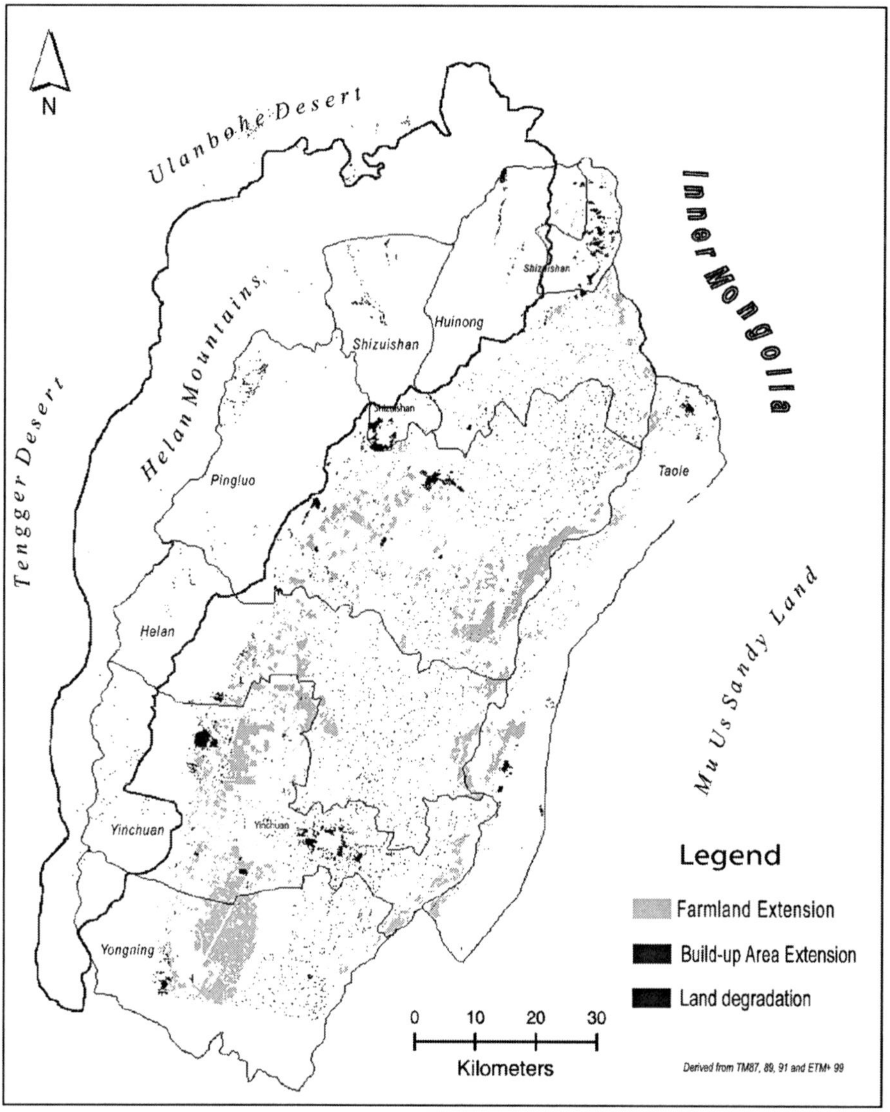

Fig. 2: Land use changes and land degradation in the region of Yinchuan, China. Derived from the differencing-threshold and change interpretation algorithm based on the Tasseled Cap Transformation (Simplified from Wu, 2004)

important change is the farmland extension, a conversion from previously sandy land and deserts into farmland (49.4% of the total change). This extension is related to the dynamics of rural population (R^2 = 0.956) and agricultural product increase (R^2 = 0.731). Land degradation, 55.8 km^2 in surface (5.5% of total change), in the forms of salinization and stone pits expansion in the Yinchuan Plain, coal dust covering, vegetation degradation and destruction of small pieces of farmland due to coal mining in the Helan Mountains, was measured. This change is associated with the industry development (R^2 = 0.708).

3.2 Case study 2

West Ordos is a part of the Ordos Plateau in China, consisting of deserts, sandy prairies and agricultural patches. Average annual rainfall is about 323 mm. Besides grazing and agriculture activity, coal mining, gas and oil exploitation have become more and more important in the recent years.

Data and processing approach are presented in Fig. 3.

3.2.1 Results

Results obtained are exposed in Fig. 4. Land degradation observed is mainly vegetation degradation due to human activity: (1) with the implementation of the Rotation Policy in the recent years, overgrazing has taken place in pastoral lands where it was permitted; (2) agricultural activity and fuel resources mining overuse underground water and lead to a decline of water-table; (3) collection of wild vegetables and herbs for medicinal use year after year has destroyed the fragile ecosystem.

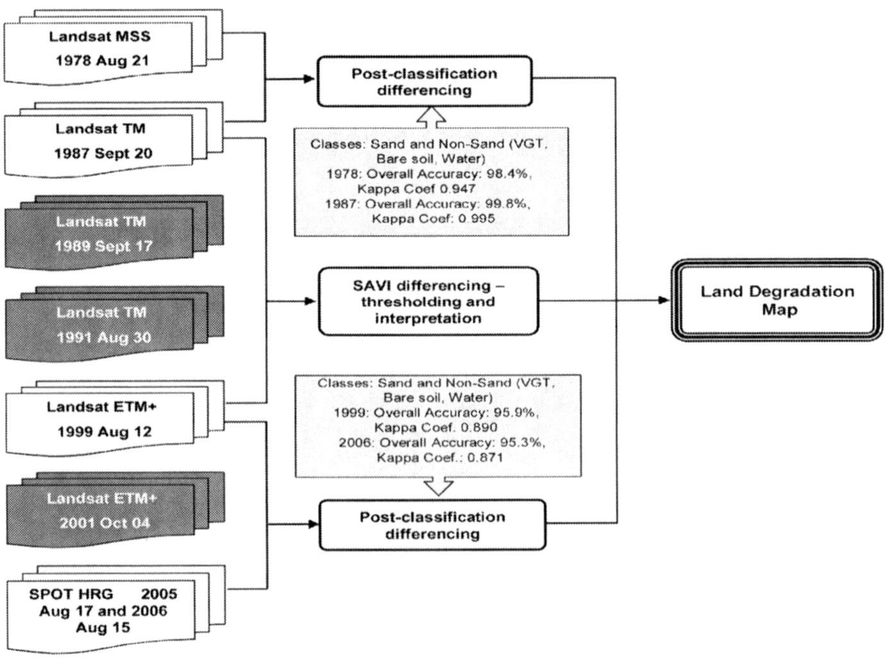

Fig. 3: Data and processing algorithm involved in the West Ordos site Note: RMS error of image-to-image correction: 0.42–0.65 pixel; images shown in grey were used as reference to check the change tendency

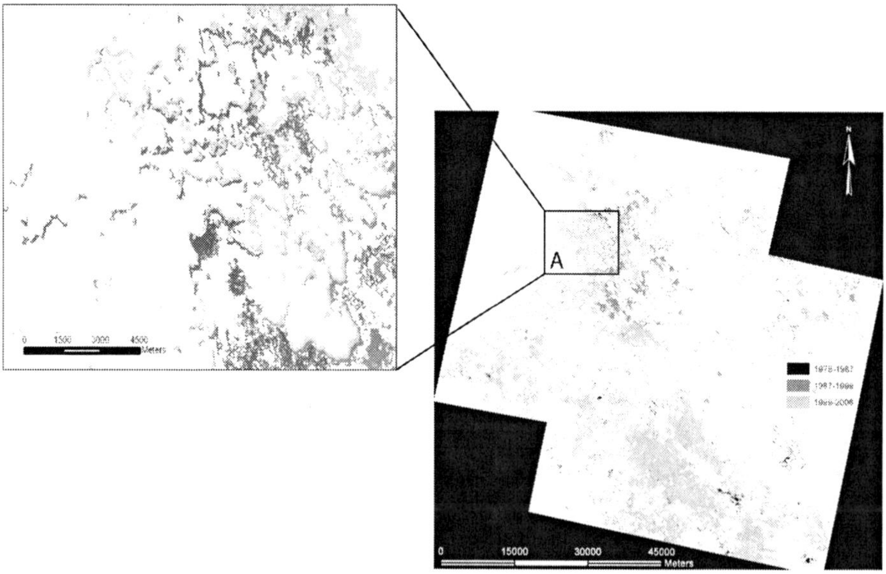

Fig. 4: Land degradation in West Ordos in the period 1978–2006. Upleft: amplification of zoom A showing grassland patches engulfed by deserts and deserts in southeastward extension

Deserts are in southeastward extension at a rate of 11–21 m/year driven by the dominant strong wind from northwest. Local government measured the sand dune movement in the field for the period March–June 2006. It is reported that the dunes moved by 1.1–4 m during this time interval. Such a velocity is comparable to what has been observed by remote sensing.

In the degraded areas, NDVI in two zooms of 15,586 pixels in total was calculated for each observed year and found a decrease trend with time. This decline seems to have some positive correlation with annual rainfall. However, the R-square value is very low ($R^2 = 0.2062$). This implies that the NDVI decrease may not have been caused by precipitation fluctuation but by other factors, as already mentioned, human activity. Detailed analysis on land degradation of this area will be seen in another paper of the author.

3.3 Case study 3

The Tarim Basin including the Taklimakan Desert is one of the hyper-arid regions of China. This case study site is located in northern margin of the basin bordered on the south front-land of the Tianshan Mountains, among which the Tarim River, the longest interior river of the country runs from West to East (see Fig. 1). The average annual rainfall is around 56–70 mm. Naturally rainfed agriculture is impossible due to aridity. The army has however, largely conducted land reclamation since 1950s. This agricultural activity has been profiting the snow water from the mountains. Land cover change and salinization has thus taken place hereafter. Li et al. (2003) and Wu (2004) have analyzed land use changes in this area.

Data used:

Corona image June 23rd, 1964, Landsat TM Oct.04th, 1994 and ETM+ July 06th, 2000. Since the TM image was obtained in October and much different in the Day of the Year from other two images, it was used as reference.

3.3.1 Procedure

Corona image is indeed a mosaic of several white and black photo slices. Therefore multispectral transformation is not applicable in this case. The post-classification differencing technique was followed. The results are illustrated in Fig. 5.

3.3.2 Results

Three major changes were observed.

1. Vegetation increase in the form of oases extension, that is, land reclamation on the pluvial plain for agriculture; it is measured that 671 km^2 of non-cultivated land has been converted into farmland from 1964 to 2000.
2. Vegetation degradation in a surface area of 175 km^2 resulted mainly from rural and urban construction and abandonment of previously agricultural land due to low productivity inside the oases and their peripheries.
3. Significant salinization as a consequence of reclamation. Inherently, the soil in the pluvial plain has already high concentration in salt. Irrigation system conducts snow-water from the mountains into the oases; dissolved salt is brought away and accumulates in the periphery areas where drainage is not developed. Salt marshes and salt ditches are thus formed. Strongly salinized lands have extended by 261 km^2 in the observed period.

Li et al. (2003) have analyzed climate changes in the period 1964–2000 and found an increase trend in annual mean temperature and annual precipitation and a decrease in annual evaporation in this area. Cleary, these observed environmental changes and degradation are not a result of rainfall decline but of anthropogenic activity.

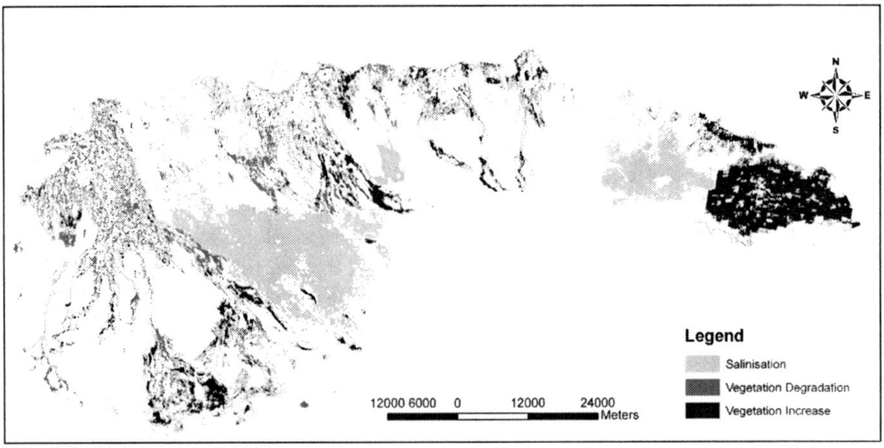

Fig. 5: Land use changes and salinization in the Middle Tarim in the period 1964–2000 (Simplified from Wu, 2003c and Li et al., 2003)

5 Summary

Based on the review of the available algorithms and the case studies, land degradation monitoring can be summarized as follows:

Remote sensing is a powerful tool and the most important information source for assessing land surface processes since it provides dynamically multi-temporal and times series information. Regional scale monitoring with coarse resolution data can target the hotspots of significant land use change or land degradation, but leaves many uncertainties that require further verification. As illustrated in these case studies, local scale observation and assessment with high-resolution data allow getting a better understanding of what has occurred and what is taking place in the identified hotspots. Multi-scale observation will be a cost-effective combination for land degradation or land use change monitoring.

Application of remote sensing in land degradation research requires great care since land degradation is a subtle and continuous process that requires to be separated from the temporal climate phenomenon like drought. Ground data and socio-economic studies are key information to explain the change or degradation mechanism and drivers. Moreover, land degradation monitoring needs to be carried out in a holistic way by linking remote sensing with human activity in order to evaluate land use management options for sustainability or rehabilitation of the natural resource base.

Bibliography

Byrne G. F., Crapper P. F., and Mayo K. K. 1980. *Monitoring land-cover change by principal components analysis of multitemporal Landsat data*, Remote Sensing of Environment, Vol. **10**, 175–184.

Caselles V. and Garcia M. J. L. 1989. *An alternative simple approach to estimate atmospheric correction in multitemporal studies*, International Journal of Remote Sensing, Vol. **10**, 1127–1134.

Chavez P. S., Jr. 1996. *Image-based atmospheric correction — revisited and improved*, Photogrammetric Engineering and Remote Sensing, Vol. **62**, n. 9, 1025–1036.

Courel M. -F., Kandel R. S., and Rasool S. I. 1984. *Surface albedo and the Sahel drought*, Nature, Vol. **307**, 528–531.

Crist E. P. and Cicone, R. C. 1984a. *Application of the Tasseled Cap concept to simulated Thematic Mapper data*, Photogrammetric Engineering and Remote Sensing, Vol. **50**, 343–352.

Crist E. P. and Cicone R. C. 1984b. *A physically-based transformation of thematic mapper data — The TM tasseled cap*, IEEE Transaction on Geoscience and Remote Sensing, Vol. **GE22**, n. 3, 256–263.

Dave J. V. 1980. *Effect of atmospheric conditions on remote sensing of a surface of non-homogeneity*, Photogrammetric Engineering and Remote Sensing, Vol. **46**, 1173.

Engvall J. L., Tubbs J. D., and Holmes Q. A. 1977. *Pattern recognition of Landsat data based on temporal trend analysis*, Remote Sensing of Environment, Vol. **6**, 303–314.

Forse B. 1989. *The myth of the marching desert*, New Scientist, Vol. **4**, 31–32.

Gao J., Zha Y., and NI S. 2001. *Assessment of the effectiveness of desertification rehabilitation measures in Yulin, North-western China using remote sensing*, International Journal of Remote Sensing, Vol. **22**, n. 18, 3783–3795.

Gordon S. 1980. *Utilizing Landsat imagery to monitor land use change: a case study in Ohio*, Remote Sensing of Environment, Vol. **9**, 189–196.

Graetz R. D., Pech R. P., and Davis A. W. 1988. *The assessment and monitoring of sparsely vegetated rangeland using calibrated Landsat data*, International Journal of Remote Sensing, Vol. **9**, n. 7, 1201–1222.

Hanan N. P., Prevost Y., Diouf A., and Diallo O. 1991. *Assessment of desertification around deep wells in the Sahel using satellite imagery*, Journal of Applied Ecology, Vol. **28**, 173–186.

Hellden U. 1988. *Desertification monitoring: is the desert encroaching?* Desertification Control Bulletin, Vol. **17**, 8–12.

Hellden U. 1991. *Desertification–time for an assessment*, Ambio, Vol. **20**, 372–383.

Hill J. 1993. *Monitoring land degradation and soil erosion in Mediterranean environments*, ITC Journal, no. 4, Netherlands, 323–331.

Hill J., Mégier J., and Mehl W. 1995. *Land degradation, soil erosion and desertification monitoring in Mediterranean ecosystems*, Remote Sensing Reviews, Vol. **12**, 107–130.

Huete A. R. 1988. *A Soil-Adjusted Vegetation Index (SAVI)*, Remote Sensing of Environment, Vol. **25**, 295–309.

Huete A. R., Liu H. Q., Batchily K., and VanLeeuwen W. 1997. *A comparison of vegetation indices global set of TM images for EOS-MODIS*, Remote Sensing of Environment, Vol. **59**, 440–451.

Hurd J. D., Wilson E. H., Lammey S., and Civco D. L. 2001. *Characterization of Forest Fragmentation and Urban Sprawl using Time Sequential Landsat Imagery*, Proceedings of 2001 ASPRS Annual Convention, St. Louis, MO.

Ingram K., Knapp E., and Robinson J. W. 1981. *Change detection technique development for improved urbanized area delineation*, Technical memorandum CSC/TM-81/6087, Computer Science Corporation, Silver Spring, MD.

Jensen J. R. and Toll D. L. 1982. *Detecting residential land use development at the urban fringe*, Photogrammetric Engineering and Remote Sensing, Vol. **48**, 629–643.

Justice C. O. 1986. *Monitoring the grasslands of semi-arid African using NOAA-AVHRR data*, International Journal of Remote Sensing, Vol. **7**, 1383–1409.

Kaufman Y. J. and Sendra C. 1988. *Algorithm for automatic atmospheric corrections to visible and near-IR satellite imagery*, International Journal of Remote Sensing, Vol. **9**, n. 8, 1357–1381.

Kaufman Y. J. and Tanré D. 1992. *Atmospherically Resistant Vegetation Index (ARVI) for EOS-MODIS*, Proceedings of IEEE IGARS'92, 261–270, New York.

Kaufman Y. J. and Tanré D. 1996. *Strategy for direct and indirect methods for correcting the aerosol effect on remote sensing: from AVHRR to EOS-MODIS*, Remote Sensing of Environment, Vol. **55**, 65–79.

Kauth R. J. and Thomas G. S. 1976. The Tasseled Cap. *A Graph Description of the Spectral-Temporal Development of Agricultural Crops as Seen by Landsat*, Proceedings of the 2nd International Symposium on Machine Processing of Remote Sensed Data, Purdue University, West Lafayette, IN, 4B41-4B51.

Koeln G. and Bissonnette J. 2000. *Cross-Correlation Analysis: Mapping Landcover Changes with a Historic Landcover Database and a Recent, Single-Date, Multispectral Image*, Proceedings 2000 ASPRS Annual Convention, Washington, DC Available at the website: http://www.glc.org/wetlands/docs/CCA_Paper.doc

Lambin E. F. 1997. *Modelling and monitoring land-cover change processes in tropical regions*, Progress in Physical Geography, Vol. **21**, no. 3, 375–393.

Lambin E. F. and Strahler A. H. 1994. *Change-vector analysis in multitemporal space: a tool to detect and categorize land cover change processes using high temporal resolution satellite data*, Remote Sensing of Environment, Vol. **48**, 231–244.

Lamprey H. F. 1975. *Report on the desert encroachment reconnaissance in north Sudan*, Oct. 21–Nov. 10, UNESCO/UNEP (mimeo), 16.

Li L., Lambin E., Wu W., and Servais M. 2003. *Land-Cover Change in Tarim Basin (1964-2000): Application of Post-Classification Change Detection Technique*, Proceedings of SPIE Vol. **4890**: Ecosystems Dynamics, Ecosystem-Society Interactions and Remote Sensing Applications for Semi-Arid and Arid Land, Editor(s): Pan, X.; Gao, W., etc.

Lillesand T. M. and Kiefer R. W. 1994. *Remote Sensing and Image Interpretation* (3th ed.), Wiley, New York.

Lupo F., Reginster I., and. Lambin E. F. 2001. *Monitoring land-cover changes in West Africa with SPOT VEGETATION: Impact of natural disasters in 1998–1999*, International Journal of Remote Sensing, Vol. **22**, 2633–2639.

Malila W. A. 1980. *Change Vector Analysis: an Approach for Detection Forest Changes with Landsat*, 1980. Proceedings of the 6th Annual Symposium on Machine Processing of Remotely Sensed Data, Purdue University, West Lafayette, 326–335.

Malingreau J. P., Tucker C. J., and Laporte, N. 1989. *AVHRR for monitoring global tropical deforestation*, International Journal of Remote Sensing, Vol. **10**, 855–867.

Mas J. F. 1999. *Monitoring land-cover changes: a comparison of change detection techniques*, International Journal of Remote Sensing, Vol. **20**, 139–152.

Mertens B. and Lambin E. F. 2000. *Land cover change trajectories in Southern Cameroon*, Annals of the Association of American Geographers, Vol. **90**, 467–494.

Price J. C. 1987. *Calibration of satellite radiometers and the comparison of vegetation indices*, Remote Sensing of Environment, Vol. **21**, 15–27.

Qi J., Chehbouni A., Huete A. R., and Kerr Y. H. 1994. *Modified soil adjusted vegetation index (MSAVI)*, Remote Sensing of Environment, Vol. **48**, 119–126.

Quattrochi D. A. and Godchild M. F. (Eds.). 1997. *Scale in Remote Sensing and GIS*, Lewis Publishers, Boca Raton, FL, p. 406.

Raffy M. 1994. *Change of scale theory: a capital challenge for space observation of Earth*, International Journal of Remote Sensing, Vol. **12**, 2353–2357, 1997.

Richards J. A. and Jia X. 1999. *Remote Sensing Digital Image Analysis – An Introduction* (3rd ed.), Springer, New York.

Richter R. 1990. *A fast atmospheric correction algorithm applied to Landsat TM images*, International Journal of Remote Sensing, Vol. **11**, no. 1, 159–166.

Rouse J. W., Hass R. H., Schell J. A., Deering, D. W., and Harlin, J. C. 1974. *Monitoring the vernal advancement of natural vegetation*, NASA/GSCF Final Report, Greenbelt, MD, 371.

Singh A. 1989. *Digital change detection techniques using remotely sensed data*, International Journal of Remote Sensing, Vol. **10**, no. 6, 989–1003.

Skole D. and Tucker C. J. 1993. *Tropical deforestation and Habitat Fragmentation in the Amazon: Satellite data from 1978 to 1988*, Science, Vol. **260**, 1905–1909.

Smith M. O., Adams J. B., and Gillespie A. R. 1990. *Reference Endmembers for Spectral Mixture Analysis*, Proceedings of the 5th Australasian Remote Sensing Conference, Perth, 331–340.

Smith O. B. and Koala S. 1999. *Desertification: Myth and Realities*, Canadian Science Writer's Annual General Meeting, May 27–29, 1999. Available at: http://www.idrc.ca/media/DesertMyths_e.html.

Swain P. H. 1976. *Land use classification and mapping by machine-assisted analysis of Landsat multispectral scanner data*, Laboratory for Applications of Remote Sensing, Information Note: 112776, Purdue University, West Lafayette, IN.

Thomas D. S. G. and Middleton N. J. 1994. *Desertification: Exploding the Myth*, Wiley, Chichester.

Tucker C. J. 1979. *Red and photographic infrared linear combinations for monitoring vegetation*, Remote Sensing of Environment, Vol. **8**, 127–150.

Tucker C. J., Justice C. O., and Prince S. D. 1986. *Monitoring the grasslands in the Sahel 1984-1985*, International Journal of Remote Sensing, Vol. **7**, 1571–1581.

Tucker C. J., Dregne H. E., and Newcomb W. W. 1991. *Expansion and contraction of Sahara Desert from 1980 to 1990*, Science, Vol. **253**, 299–301.

Woodcock C. E. and Strahler A. 1987. *The factor of scale in remote sensing,* Remote Sensing of Environment, Vol. **21**, 311–332.

Wu W. 2003a. *Application de la géomatique au suivi de la dynamique environnementale en zones arides, PhD thesis*, Université de Paris 1.

Wu W. 2003b. *Evaluation of land use and cover change in north Shaanxi, China*, Photo-Interpretation, Vol. **39**, n. 2, 15–29, plates 35–45.

Wu W. 2003c. *Land Use and Land Cover Changes in the Critical Areas in Northwestern China*, Proceedings of the SPIE on Remote Sensing for Agriculture, Ecosystems, and Hydrology V, Vol. **5232**, 245–256, Edited by O. Manfred, D. Guido (published in Feb. 2004).

Wu W. 2004. *Environmental Change Monitoring — A Case Study in the Region of Yinchuan, Ningxia, China*, Proceedings of Map Asia 2004, Beijing, Aug. 26–29, 2004.

Wu W. 2007. *Coastline evolution monitoring and estimation — a case study in the region of Nouakchott, Mauritania*, International Journal of Remote Sensing, Vol. **28**, n. 24, 5461–5484.

Wu W., Lambin E. F., and Courel M.-F. 2002. *Land Use and Cover Change Detection and Modeling for North Ningxia, China*, Proceedings of Map Asia 2002, Bangkok, Thailand, Aug. 6–9, 2002. Available at: http://www.gisdevelopment.net/application/environment/overview/envo0008.htm

Wu W., Zucca C., and Enne G. 2005. *Land Degradation Monitoring in the Ordos Region, China*, Proceedings of Remote Sensing and Geoinformation Processing in the Assessment and Monitoring of Land Degradation and Desertification (RGLDD), p. 618–625, Sept. 7–9, 2005, University of Trier, Germany.

Young S. S. and Wang C. Y. 2001. *Land-cover change analysis of China using global-scale Pathfinder AVHRR Land cover (PAL) data, 1982-92*, International Journal of Remote Sensing, Vol. **22**, n. 8, 1457–1477.

PARTICIPATORY MONITORING AND EVALUATION OF A PROJECT TO COMBAT DESERTIFICATION IN DRYLANDS (*CASE STUDY IN CENTRE WESTERN TUNISIA*)

BOUBAKER RADDAOUI
Office de l'Elevage et des Paturages (OEP), Kasserine, Ira – Médenine, Tunisie
Raddaoui04@yahoo.fr

Abstract

The success of projects to combat desertification is measured through their contribution to the achievement of outcomes. Monitoring and evaluation activities must increase their focus on outcomes by shifting towards better measurement of performance and more systematic monitoring. Such activities must be inspired by an organizational culture of learning, transparency and accountability.

A case study of participatory monitoring and evaluation is presented in a pilot project to combat desertification that involves the local agro-pastoral communities and is implemented by OEP (Office de l'Elevage et des Paturages) and NRD (Nucleo Ricerca Desertificazione – University of Sassari), in Feriana, Governorate of Kasserine in centre western Tunisia. It aims to recover degraded rangelands by planting fodder shrubs, as well as complementary activities involving public awareness and targeted to training local users. The approach used was based on participatory monitoring tools, associated with remote sensing data and GIS, and specific indicators tests developed by OSS-SMAP/CE oriented to monitor the implementation of the National Action Plan (NAP) to combat desertification, at the sub-national and local level in Tunisia, and to evaluate its impact on the environment and natural resources management capacity of local communities.

Appropriate indicators were adopted to check both the biophysical outcomes and their social impact. Some of the indicators used as remote sensing parameters, which were assessed using an 'expert' approach, while others required a direct interaction with local breeders and farmers.

Preliminary results showed that the plantation of fodder shrubs has a positive impact on soil stability, wind erosion and water runoff control, but plantation success is strictly related to climatic conditions and the appropriate involvement of the local farmers.

The indicator 'fodder supply in period of pastoral deficit' has proved its reliability and credibility to the extent that it accurately confirms that the impact of shrub plantations cannot be determined in cases where maintenance is no longer assured and the grazing ban has been lifted. The parameter 'appropriation and charge holder level' accurately indicates plantation durability and restoration techniques effectiveness. Local management capacities were measured in terms of compromises, project activities, internal conflicts, and the identification of priorities and needs of

A. Marini and M. Talbi (eds.), *Desertification and Risk Analysis Using High and Medium Resolution Satellite Data*,

the local population, and inquiries were made on the socio-professional organization's reliability and representativeness. It revealed that the professional structure created by the project is still unable to ensure sufficient assistance to the population and to manage the divergent project objectives and expectations. Participatory monitoring and the evaluation of the socio-organizational capacity for managing natural resources, for collaboration, and for conflict-solving emerged as an effective tool towards strengthening local participant's capacity for land restoration techniques. It also generated some practical recommendations to be considered while implementing other development projects dealing with combating desertification.

Keywords: Participatory monitoring and evaluation, combating desertification projects, local people organization, rangeland improvement, grassroots indicators

1 Introduction

Increasing awareness among project planners and managers of the complexity and unpredictability of development projects, and the knowledge that the same approach can lead to different outcomes depending on the context, has led to greater orientation in terms of the process of development projects in the last decade. This requires a good system to obtain and analyze information based on experience during the course of project implementation so as to recognize weaknesses, adapt methods, correct the procedure if necessary, and respond more quickly to unexpected events (Mosse et al., 1998). Furthermore, local needs within the community change over time and development projects have to respond to these changes.

Monitoring and evaluating a development project, especially one claiming to operate in a participatory mode to combat desertification, is not simply a matter of looking at project activities. The relationship between the project and its partners, including the beneficiaries, must also be closely looked at. In such an approach, the partners get to better understand each other, avoiding unrealistic expectations and disappointments, and make project success more likely (Reckers, 1997). Participatory monitoring and evaluation (PM&E) is an integral part of local capacity building and institutional development. It can create a feeling of ownership among all partners and can help local people managing their own affairs better and thereby increasing the likelihood that project-supported activities will continue after the project will end (Wolfgang Bayer, Ann Waters-Bayer, 2002).

On the other hand, it could be argued that local resource users have the right to know the amount of funds allocated for different project components and its justification. This could be examined during periodic evaluations. In contrast, if a project is supporting the establishment of a private service or a cooperative or any socio-professional structure, financial monitoring should clearly be done in cooperation with both local people and the project partners, as this activity helps to strengthen local capacities and continues to supports the initiative long after the end of the project.

Furthermore, in spite of knowledge and data accumulation, and the significant progress in terms of understanding the biophysical processes underlying land degradation, very few approaches to monitoring and evaluating projects implemented to combat desertification have been undertaken which evaluates project success and supports the decision-making process in Tunisia (UNEP, 2001). Such approaches are particularly needed in the vulnerable arid regions where the problems of land degradation are most critical and where there are high inputs invested by governments to carry out desertification mitigation activities.

This work is carried out in parallel of the trend observed among United Nations and bilateral development agencies and the European community and Non Governmental Organizations. It provides an opportunity to present relevant experience and results obtained by participatory monitoring and project evaluation to combat desertification implemented in semi-arid zones in Tunisia.

2 Methodological approach

2.1 Study area

The area considered it is located in the sector of Skhirat (Delegation of Fariana, Governorate of Kasserine) that lies south of the western ridge and in the piedmont of Chaambi Mountain (Fig. 1).

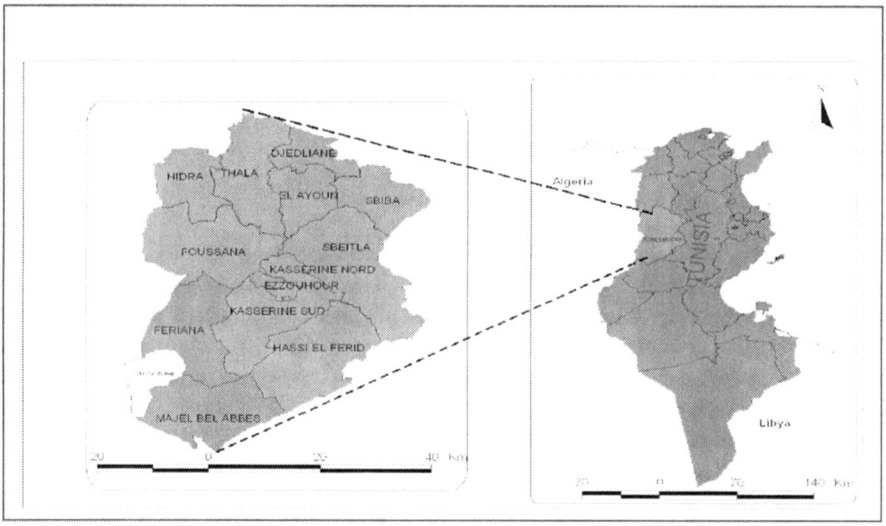

Fig. 1: Geographical location of the project

The target area adequately represents some of the main land degradation and desertification issues typical of Tunisian territory and represents a significant example of the dramatic extent of land degradation reached by the rangelands in the recent years.

2.2 Project implemented in the study area (Skhirat)

The development project titled 'Demonstrations on strategies to combat desertification in arid lands with the direct involvement of local agro-pastoral communities in west central Tunisia' aims to contribute to the restoration of vegetation cover with drought resistant perennial species in highly degraded extensive pasture land, as well as the mitigation of desertification processes through rangeland improvement and a number of rehabilitation techniques.

These objectives were consistent with the goals of the 10th five-years Plan for Socio-Economic Development (2002–2006) as well as with the priorities of the country assistance strategy and its National Action Plan to combat desertification. The afore-mentioned objectives are to be achieved through a series of activities that are expected to obtain the following results:

Direct restoration and intervention on the land by using perennial fodder shrubs and forage species.

Training activities and specific studies to strengthen local capacities and knowledge on restoration techniques.

Complementary demonstration, public awareness and information dissemination activities at local and national levels, as well as the promotion of the international debate on desertification mitigation measures suitable for North African countries.

2.3 Data collection method

Appropriate participatory tools were applied in gathering spatial information: community sketch map and transects. The project activities, baselines, inputs and outputs, and targets are collected from the annual project reports and confirmed in the field by the project beneficiaries.

2.4 Monitoring and evaluation tools applied

To evaluate the impact of project activities, we have tested and applied the indicators catalogue conceived by OSS-SMAP/CE-Tunisie (2004) with reference to the corresponding project activities and the conceptualization approach proposed by Talbi (2005). The data collected are treated in a database called ‘Effort and Impact’, which was developed as a platform for the various partners involved in implementing the Tunisian NAP to combat desertification (OSS-SMAP/CE-Tunisie, 2005). Project activities were monitored through the use of critical indices (NDVI), high-resolution satellite imagery, and imagery from different sources taken in varying times to detect and discriminate changes attributed to the project.

3 Evaluation of the achieved project activities

The evaluation concerns the impact of two achieved activities: ‘Fodder shrubs plantation and pastoral enrichment’ and ‘Socio-professional organization’, which represents fundamental actions to mitigate land degradation and to ensure land users organization and which are expected to assume responsibility for operating and maintaining facilities at the local level and feedback while enriching decisions made at policy level.

Fodder shrubs plantation and pastoral enrichment:

This Development Activity (DA), frequently used in rural development projects, responds to drought effects mitigation by the creation of fodder reserves, pastoral enrichment and soil protection against erosion. The species used in this study include: *Acacia cyanophylla*, *Opuntia ficus* and *Atriplex nummularia*. The appropriate indicators, with their correspondent parameters, are shown below:

Indicator 1: Fodder supply in period of pastoral deficit
Parameter 1: exploitation of the plantation level
Parameter 2: the onset of the state of spontaneous vegetation
Indicator 2: Soil stability
Parameter 1: state of erosion
Indicator 3: Plantation durability
Parameter 1: level of ownership or level of responsibility of the plantations

The methodology consists of evaluating the DA's impact using the above indicators, as well as the assessment of the productivity of fodder shrubs, their successful growth, and the onset of spontaneous production of vegetation cover.

Socio-professional organization of land users:

The expected impact of this activity was seen at mean term, in the five-year evaluation. In the short term, with respect to annual monitoring and evaluation, it should be verified whether the conditions required to reach the expected impact have been established.

Indicator 1: Reliability of the socio-professional organization (SPO)
Parameter 1: compromises management capacity
Parameter 2: DA's management capacity
Parameter 3: internal conflict management capacity
Indicator 2: Socio-professional organization representativeness
Parameter 1: priorities and needs identification capacity
Parameter 2: incitement measures, valorization capacity

Evaluation of sustainability of rehabilitation activities

According to Abbot and Guijt (1998), project sustainability can be measured using indicators that reflect the awareness rising of local users. Such indicators were adopted in this study to evaluate the impact of the training sessions and public awareness workshops on local-user groups. These groups were determined by analysis of the lists of project beneficiaries, available in the annual project reports, and triangulated and validated with relevant data offered by the socio-professional organization in the project area.

Indicator 1: *Percentage of people participating in some form of land rehabilitation.*

Indicator 2: *Percentage of rural adults involved in the project activities.*

Indicator 3: *Number of beneficiaries who carry out plantation maintenance by themselves.*

4 Results and discussion

4.1 Community sketch map of Skhirat

The Community Sketch Map showed where resources, problems, opportunities and project activities for remediation are located (Fig. 2). The mapping also identified areas with specific problems (such as water-shortage in Eltaifia, Ouled Abid and Ouled Brahim, erosion and gulley formation in the south, overgrazing due to overstocking everywhere in the area) or potential for improved production (e.g. predominant agricultural and livestock existing in the area for income-generation), and the project response to these issues.

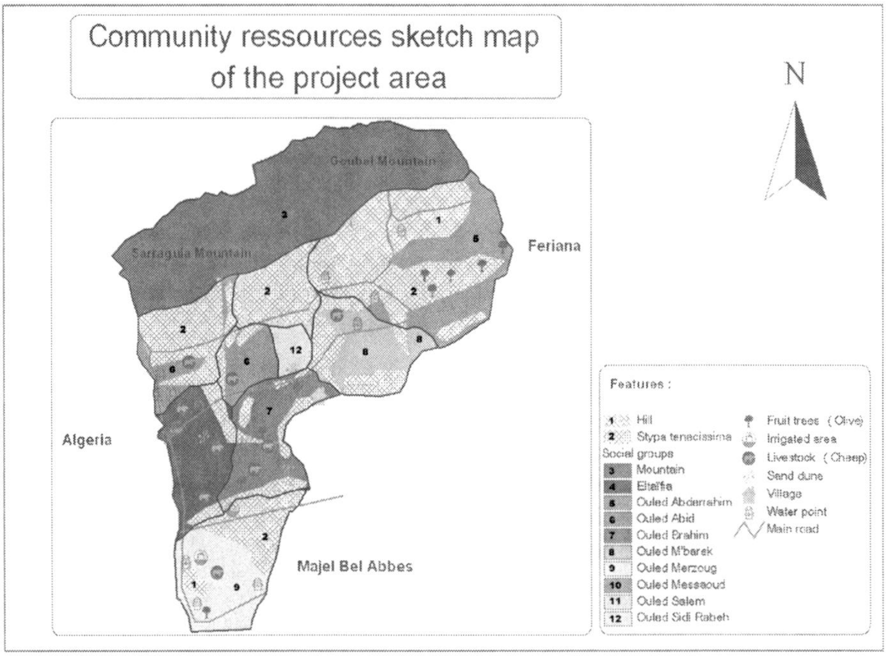

Fig. 2: Community resources sketch map of the project area – Normalized Difference Vegetation Index (NDVI)

It was not possible to distinguish between dryland farming and the rangeland – whether it is rehabilitated or not, poor, fair or good as the reflectance between rainfed cropland and rangelands in the Mountain and Hills is mostly the same (Fig. 3).

4.2 Change detection using high resolution imagery

The Quick bird satellite image (0.60 m of resolution) clearly distinguishes the palisades of palm leaves, constructed in the project area to stop sand encroachment, and associated with the fodder shrubs plantations. The image points out that the project focused on rangeland improvement where native species dominated and intervened in vulnerable sites to establish fodder reserves; protecting these plantations will control overgrazing, improve soil stability and prevent degradation.

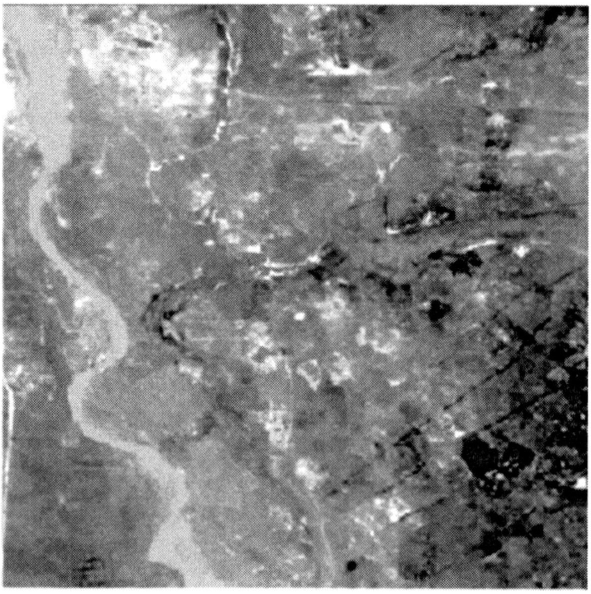

Fig. 3: NDVI QB August 1st, 2005

4.3 Change discrimination using different satellite imagery

The area units I and II, shown on the Aster image, represent a sample of degraded rangeland with the bare soil subjected to water and wind erosion (Fig. 4). On the Quick bird image, the same area units were rehabilitated by the project and show a different reflectance. The ground truth has confirmed that the units are restored using *Acacia cyanophylla* and land cover is improving.

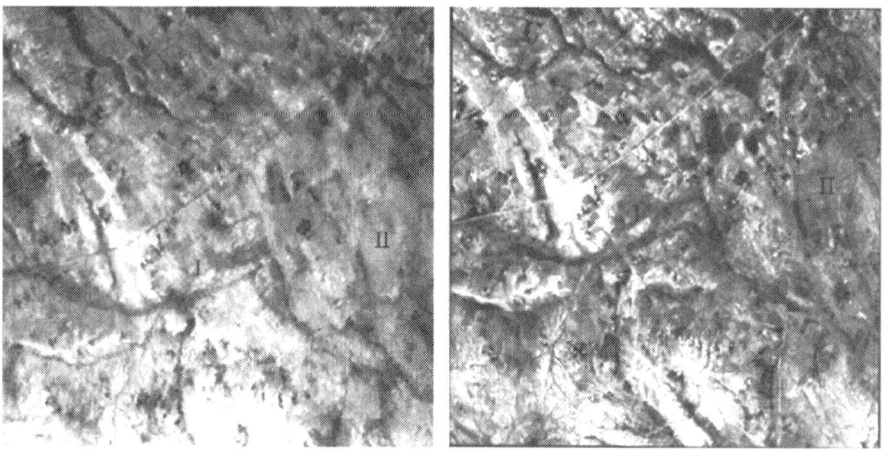

Fig. 4: Discrimination change using high resolution image

The desired state can be defined by targets that include a rehabilitation of rangeland by fodder shrub plantations to decrease the area of low infrared signal.

4.4 Evaluation of DA: fodder shrub plantation and pastoral enrichment

Soil stability is assured wherever the plantation growth has proved successful and when the grazing ban has been genuinely respected. Plantations on the other sites seem not to be sustained and expressed weak impact as regards to soil stability (Fig. 5). Consequently, the soil is not well covered and is subjected to erosion and degradation. These results indicate that the rehabilitation techniques, using fodder shrubs, applied to improve degraded rangeland do not achieve the same impact in the whole study area as when targeted in the project objectives.

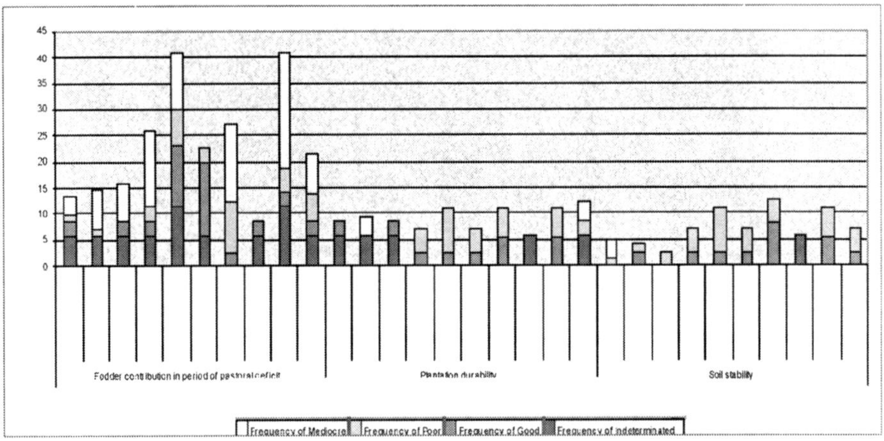

Fig. 5: Impact of the DA "Plantation of fodder shrubs and pastoral enrichment"

4.5 Productivity and success rate of Shrubs introduced

For the new plantations, two cladodes per individual will not ensure its protection against frost during winter or the prevailing dry and hot wind in summer. This explains the low growth success in the cacti introduced in 2005. The Atriplex has shown a high success level compared to Acacia due to its notable drought resistance and salt tolerance. The low success rate of Acacia could be attributed to irrigation water shortage, the cost of maintenance, and the fact that fruit trees with their associated cash flow compete more than shrub plants. Cactus plantations provide important green biomass on the third year following plantation. They can reach 2.25 t/ha if the plantations are well maintained and/or fertilized. Growth success, expressed in terms of new produced cladodes, is highly affected. Plantations of 2003 have high growth success and therefore high production since they escaped frost effects.

The onset spontaneous vegetation cover and production:

The production of spontaneous vegetation inside parcels made up of cacti can reach 4.88 t/ha, and the cover is more important than those with Acacia or Atriplex

(Figs. 6, 8). This is due to the high vegetative biomass of annual species. While biomass production of spontaneous vegetation inside the parcels of Atriplex do not exceed 4.2 t/ha within three years, the nature of the composition of the onset vegetation, mostly based on perennial species, provides a sign of sustainability (Fig. 7).

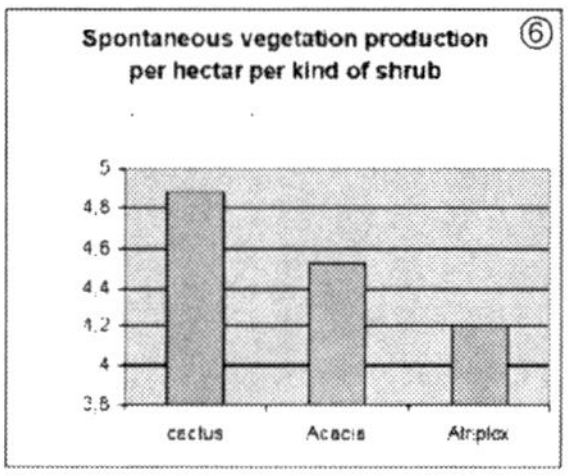

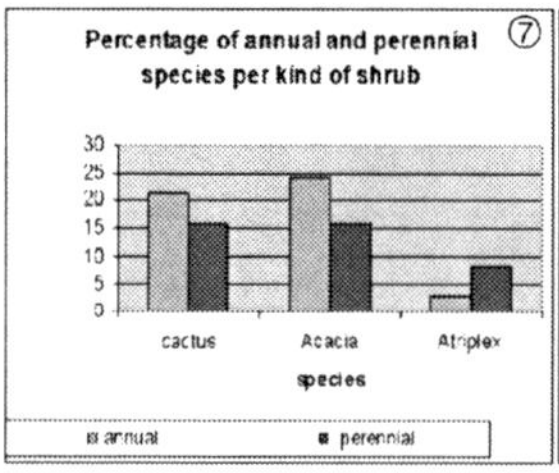

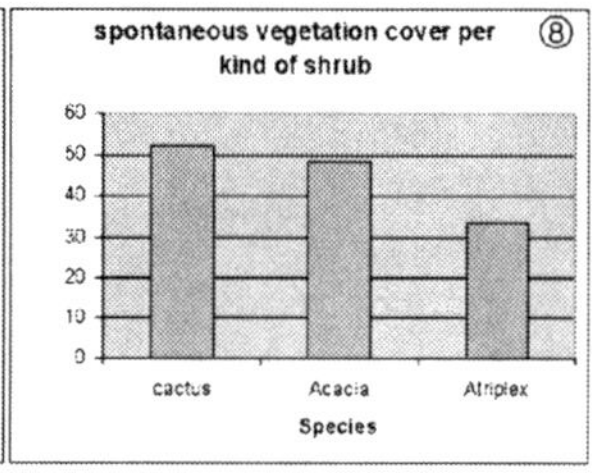

Fig. 6: Percentage of vegetation
Fig. 7: Percentage of annual and perennial species
Fig. 8: Spontaneous vegetation production per kind of shrub

Measured parameters analysis (Reliability and Representativeness) shows that some of the social groups (Ouled Abderrahim) concerned by the project are not represented. The SPO's reliability, although found to be weak (10%), is the same over the whole area. The situation is different for Ouled Merzoug, since they had an effective participation and a background due to their participation in stakeholder organization acquired from previous projects (Fig. 9).

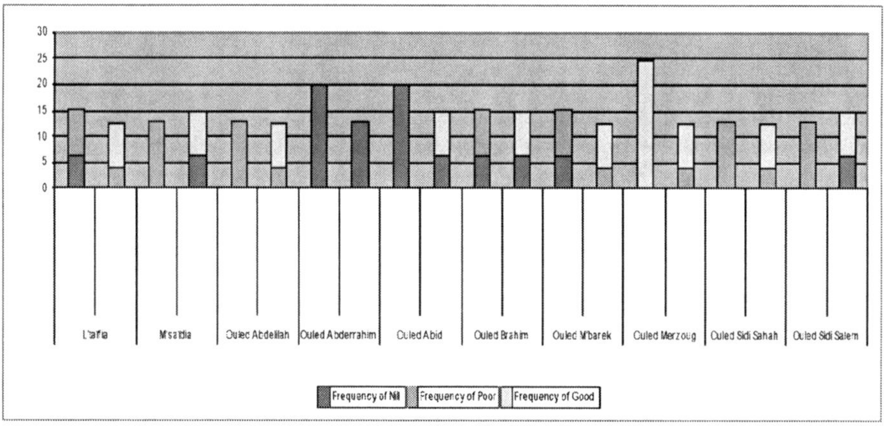

Fig. 9: Socio-professional organization impact

The recently created SPO does not have the sufficient financial means or trained staff to ensure adequate assistance to the population and in order to manage divergent objectives and expectations. Nevertheless, this is not a sign of project failure, as some specific training sessions oriented to both local users and SPO's members are ongoing. Moreover, in the future, if one were to look again at the indicators at the end of the project they would reveal substantial progress even beyond expectations.

4.6 Evaluation of project sustainability

Indicator 1 showed that the percentage of people participating in rehabilitation activities dropped from 40% in the first year to 20% in the second year, which subsequently increased to reach 32% in the third year (Fig. 10). The decrease is attributed to the prolonged drought registered in 2004.

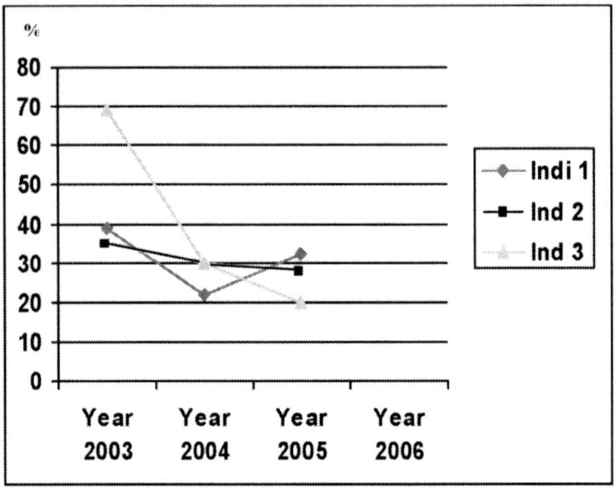

Fig. 10: Project sustainability measurement

Indicator n. 2 proves that there is a high percentage of youth participating in land rehabilitation.

Indicator n. 3 revealed that 20% of beneficiaries are aware of the importance to maintain damaged plots at the end of the third year (2005). This progress can increase even beyond expectations in the near future, if intensive training and public awareness associated with close supervision is addressed to local people.

4.7 Performance of the project

The basic objectives of the project conformed to priorities of the Tunisian National Action Plan to combat desertification and the evolving policies of the Socio-Economic Development plan.

In most sites, the sustainable regeneration of degraded rangeland already rehabilitated by the project is being achieved (Fig. 11). The lack of beneficiaries' awareness regarding desertification risk and the limited financial capacities to support maintenance and widespread activities to prevent degradation, constitute a real hindrance to attaining project objectives.

4.8 Lessons learned

One of the lessons learned from this study, and an aspect that is usually not foreseen in project planning, is that monitoring and evaluation activities could contribute to

Fig. 11: Rangeland rehabilitation using Opuntia ficus

strengthening grassroots organizations. This effect, which is largely achieved through participatory, monitoring and evaluation in the study area, was evaluated as an unforeseen but important outcome of the overall project.

Conclusion and recommendations:

Combating desertification through rangeland rehabilitation measures with the involvement of trained local agro-pastoral communities has proved its success in sites where the grazing ban has been assiduously respected. However, fodder shrub species used to improve degraded rangeland (*Opuntia ficus*, *Acacia cyanophylla*, *Atriplex nummularia*) did not show the same growth success and impact in the whole study area, as targeted in the project objectives. This relies on the participation of beneficiaries and their initiatives to carry out maintenance activities, as well as the resistance capacity of the used species to harsh climatic conditions, especially prolonged drought and frost. Project activities need to be selectively targeted if the most disadvantaged groups are to benefit, and this should receive greater attention in later projects. Moreover, the demand for training should precede its provision, and the preferences of group members should be respected.

Bibliography

Abbot J. and Guijt I. 1998. *Changing views and changes: anticipatory approaches to monitoring the environment,* Discussion paper n. 2, International Institute for Environment and Development (IIED), London. 96 pp.

Mosse D., Farrington J. and Rew A. 1998. *Development as process: concepts and methods for working with complexity.* Routledge, London. 202 pp.

OSS-SMAP/CE – Tunisie. 2004. *Conception d'un dispositif opérationnel de suivi-évaluation au niveau sub-national et local en Tunisie: Mise en œuvre opérationnelle.* L'Observatoire du Sahara et du Sahel, Tunis, juin 2004. 20 pp.

OSS-SMAP/CE – Tunisie. 2005. *Plaquette base de données 'Efforts et Impacts',* L'Observatoire du Sahara et du Sahel, Tunis, mai 2005.

Reckers U. 1997. *Participatory project evaluation: letting local people have their say.* Development in Practice, n. 7. No. 3. pp. 12–14.

Talbi M. 2005. L'opérationnalisation du dispositif de suivi évaluation du PANLCD en Tunisie. Document de synthèse. Tunis, septembre 2005. 23 pp.

UNEP. 2001. *Plan bleu. Centre d'activité régionale; Tunisie.* Nouvelle Série, Problématique et Politiques d'Environnement et de Développement Durable. 90 pp.

REMOTE SENSING TECHNOLOGY APPLICATIONS FOR DESERTIFICATION MAPPING: A CASE STUDY, OUDIA AREA, TUNISIA

MOHAMED OSMAN ARNOUS[1,2], MOHAMED AHMED SIDI CHEIKH[2], BEN ZAIED MONGI[3], RIADH ALIOUT[4], RABAH HADJ-ALI[4], YASSINE GAROUNI[4], SALEH SID[4], ALBERTO MARINI[5], AND FRANCESCO MUNTONI[5]

[1] *Suez Canal University, Faculty of Science, Geology Department, Ismailia, Egypt*
[2] *Parc National du Banc d'Arguin, Muretania*
[3] *Institut des Regions Arides, Tunisia*
[4] *University Mentouri Constantine, Algeria*
[5] *University of Cagliari, Earth Science Department, Italy*

Abstract

Ecosystems in semi-arid regions appear to be undergoing degradation processes commonly described as desertification. Desertification has been widely recognized as one of the most serious environmental problems in world, although its concept and causes are still debated in scientific communities. The semi-arid region in North Africa is ideal for investigating the processes and mechanisms of desertification due to its very sensitive ecosystem. The desertification constitutes one of the major concerns of the countries of the Sahel and North Africa. This phenomenon is responsible for the degradation of natural habitat and for arable land disappearance. Using the Sandy land of Choott El-Garssa, a former dormant dunes and wetlands environment, we monitored the variation of the desertified land and the wetlands in order to assess the climatic and human impacts on the processes of desertification. The research methods included digital image processing of high resolution of satellite data and field verification. Remote sensing data shows that the wetlands were larger in Oudia area. The variation of the extension of mobile dunes is in inverse proportion to that of wetlands. The main process of desertification is the reactivation of dormant sand dunes. The changes of wetlands and desertified lands correlate considerably well with the variations of local precipitation and temperature, suggesting that climate might be the key factor triggering desertification in the Oudia area. In other sandy lands of Oudia area, however, human activities might be the key factor causing desertification and land degradation. With the development of remote sensing techniques, it became possible to study this phenomenon through several scales and to analyze the interaction between the various elements of the environment in relation with the dynamics of soil and the human activity. This study was an attempt to use high-resolution satellite data Quick Bird for mapping desertification in the zone of Tozeur. The example selected for this paper is the sector of Oudia located about 12 km North West of Tozeur. The methodology used is the simultaneous approach of geomorphological and digital analysis of high-resolution satellite data for defining and mapping the desertification landscape. Visual interpretation combined with supervised classification using training site tests allowed us to perform a precise

A. Marini and M. Talbi (eds.), *Desertification and Risk Analysis Using High and Medium Resolution Satellite Data,*

identification and led us to an accurate mapping desertification sensibility. As a result, it appears that Oudia is threatened by the encroachment of sand dunes and a significant wind activity. In conclusion, the study strongly recommends the use of these new techniques of remote sensing in desertification assessment and mapping.

Keywords: Desertification, land degradation, Remote Sensing, GIS, Quick bird, Tozeur

1 Introduction

Ecosystems in semi-arid regions appear to be undergoing degradation processes commonly described as desertification. Desertification represents one of the most threatening environmental hazards due to the large amount of people and land at risk (Clarke and Noin, 1998). Although numbers vary across the literature, it is estimated that desertification directly affects one-third of the Earth's land and 250 million people, most of them in developing countries; even a higher number could be potentially affected (Clarke and Noin, 1998; Murray, 1999; UNCCD, 2004a).

Desertification is the label for land degradation in arid, semiarid, and dry sub-humid areas, collectively called drylands. Land degradation is defined as the 'reduction or loss of the biological or economic productivity and complexity of agricultural land (including rainfed cropland, irrigated cropland, range and pasture), forests and woodlands' (United Nations, 1994). The term implies a negative change with respect to previous environmental conditions, often implicitly considered as the 'normal' or desirable state of affairs in the ecosystem in terms of its use and management (Blaikie and Brookfield, 1987).

Desertification is the outcome of different processes, including climatic variation and human activities (United Nations, 1994). Regarding climatic variation, moisture deficiency or aridity is the most common climatic criteria used for the definition and classification of drylands, and is caused when water losses (evaporation and transpiration, referred collectively as evapotranspiration) exceed water inputs (precipitation) (Strahler and Strahler, 1992). Low and irregular rainfall, high evapotranspiration, and cyclical droughts contribute to this deficiency.

Human activities leading to land degradation in drylands are related to land uses and habitation patterns, and include: (a) soil erosion caused by wind and/or water; (b) deterioration of the physical, chemical and biological, or economic properties of soil; and (c) long-term loss of natural vegetation (United Nations, 1994).

An immediate consequence of the United Nations' definition of desertification is that its understanding requires taking into account two different but related aspects: (1) the natural dynamics, average characteristics, and variations of the physical and biological components of arid and semiarid environments, their evolution over time, and their spatial diversity; and (2) how populations have historically used the natural resources in a given area, and what kind of environmental problems may stem from different forms of land use and management.

Desertification has been defined as "land degradation in drylands resulting from various factors, including climatic variations and human activities" (UNCED, 1992). As used here, the term "land" includes soil, local water resources, as well as natural vegetation or crops. "Degradation" implies reduction of resource potential by one or a combination of processes acting on the land. Overgrazing has been recognized as a major factor of desertification in many arid regions of the world, and the effects of livestock on vascular plants have been widely studied. Areas with minimal or no anthropogenic activity are usually used as reference areas for assessing degradation processes. At the same time, the processes occurring in protected and/or unused areas should also be of interest for scientists dealing with desertification.

The desertification mapping by remote sensing has an important role in these last years. Therefore, in the current work, we seek contributing to the knowledge of the environmental and social dynamics of drylands, by exploring environmental hazards related to aridity and desertification in Oudi area, an arid area located in the province of Chott El-Garssa the South of Tunisia. The general objective is twofold. We intend to address environmental change, identifying and locating land degradation processes. Also, we intend to establish the relation between these processes, the natural dynamics of the ecosystem, and human activities and settlement.

This Remote Sensing cartography serves to redefine the current limits of zones affected by desertification and as source data to decision-makers for the implementation of more effective means of fight against the desertification. Monitoring by satellite images allows, in particular, to follow the different modifications occurred to the soil.

The South of Tunisia and particularly the depressions of Chott El-Garssa are strongly threatened by desertification amplified by human pressure, demography, overgrazing and urbanization. Today the surface of the zones affected by the desertification is in progression. The human pressure makes itself more and more intense. This pressure appears by the anarchical exploitation of the space.

During the period from 9th till 19th December 2006, an international workshop on the desertification mapping by the use of remote sensing was organized in the city of Tozeur (Tunisia). The participants benefited from technical help given by several specialists in geomorphology of arid areas and several excursions on field were organized.

This paper attempts to summarize the work achieved by the team II on the sector of Oudia. The zone concerned by this study is the sector of Oudia. In spite of the apparent state of good sand fixing, this zone is threatened by the formation of barkhanic dunes, which constitute the most mobile shapes of dunes.

The maps that we propose in this report have been achieved from satellite data at very high resolution Quick Bird and Landsat (ETM +7) by using simultaneously the visual interpretation and the digital processing. Our approach was based on the use of Quick Bird satellite data to build up the following thematic maps: the morpho-dynamic map; the land cover map; the sand cover map.

The present cartographic survey will permit to give the present limits of the geomorphological structures, the description of the plant formations, the use of soil and the state of sensitization to the desertification, while using visual interpretation and digital analysis. These maps should be used as source data to the operations of planning and definition of a global strategy allowing the government and the authorities concerned to make measurements to minimize the impact on soil.

2 Study area

The study area is located in the Southwest of Tunisia (Fig. 1). It is about 400 km from the capital Tunis. The zone of study is situated in the neighborhood of Chot El-Garssa and Chott El-Jerid, which constitute a depression area. It is surrounded by important geological formations. Located at the north of Tozeur, Chott El Gharsa constitutes a real closed depression which spreads out on an about 1,350 km. Lengthened on approximately 50 km and having altitudes lower than the sea level (between -5 to -25 m). This depression represents the basic level of the whole of the hydrographic network, which drains the chains of the southern Atlas and the Western part of the low southern plains (Talbi et al., 2006). It is among the places that made think to dig an active channel like Suez from Gabèse to the Sahara Algerian.

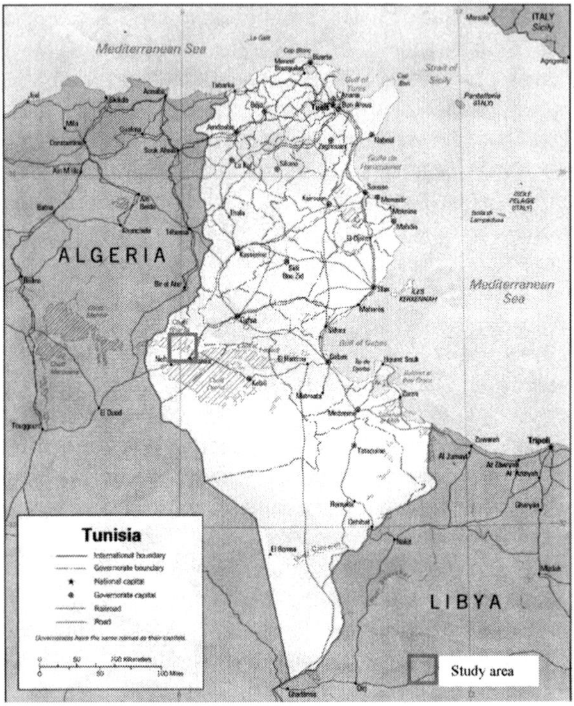

Fig. 1: Study area in Central Tunisia

The annual precipitations are between 50 and 100 mm (Talbi et al., 2006). A very contrasted thermal regime and very important annual and diurnal amplitudes show the very pronounced continental character of the region of Tozeur. The average temperature is about 22°C (Talbi et al., 2006).

We register an enormous evaporation climate power. This high evaporation, combined in a weak rainfall, causes an annual strongly deficit balance. The zone of study is subjected to important flows of wind drained by a vast network of wind corridors through the chains of mountain bordering the chotts, and arranged according to a west-east axis. The zone of study is characterized by very extended and localized soils. The scarcity of soils in this region finds its explanation in the revealed aridity, the very scattered plant cover, moreover to the influence on the quality of the soils by the depressions closed during the Quaternary (Talbi et al., 2006). The study area is distinguished by sparse low vegetation, which is usually very degraded by the overgrazing.

3 Materials and methods

3.1 Remote sensing data resources

The remote sensing data utilized in this research are Quick bird and Landsat ETM+ 7.

- Two Quick bird images were acquired in February 2006, the first one is panchromatic of 60 cm resolution and the second one is multi-spectral of 4 m resolution containing 4 bands. These data come from the laboratory of TeleGIS of the University of Cagliari. These data were used because in Tunisia and in joining areas the maximum growth of vegetation takes place in the end of winter and spring.
- Landsat ETM+7 data set (Row 36/Path 192 of the WRS-2), dated 6th October 2000, The panchromatic band of the ETM+ at 15 m spatial resolution was re-sampled to 30 m by a bi-cubic convolution method. This image comes from the Web site of Global Land Cover Facility. The mosaic image was acquired between November 1999 and June 2001.

Initially, we carried out the fusion of two images Quick Bird (60 cm and 4 m). For that purpose, we used the software ERDAS Imagine. Figure 2 shows the Quick Bird image resulting from of the fusion operation. It has a 60 cm of resolution. These data are geo-referenced to UTM coordinate system as UTM/WGS 84, zone 32N.

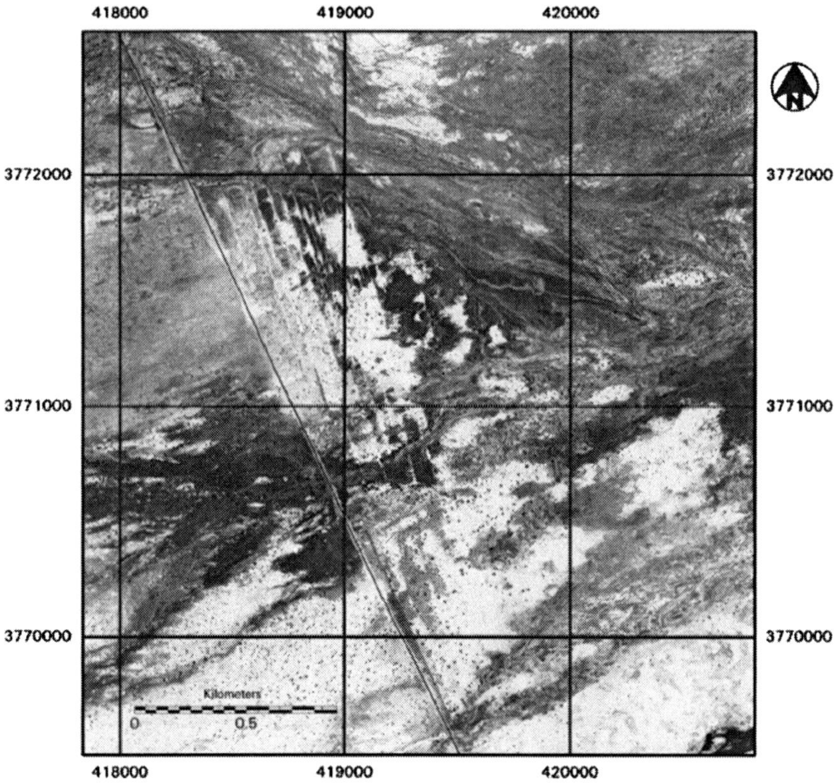

Fig. 2: Quick Bird image taken on February 2006

3.2 Methods of data collection

The fieldwork and verification were carried out as one of the most important steps. The identification and information extraction from the enhanced satellite images, of a series of representative training sites, allow us to recognize the objects on ground. According to spectral signatures, texture, pattern and tone of false color composite the training site on screen were selected. The spectral signatures of the training samples of the study area are Sabkha, sand dune, barkhan and salt crust.

Over a period of three days data about soil and plant have been collected with the aid of the Global Positioning System (GPS) and enhanced Quick Bird images and are used as base map. Moreover, in the field works a vehicle 4 X 4 and a digital camera are used. A Tunisian expert accompanied teamwork (Nabil Gasmi), who has a very good knowledge of the zone of the survey.

During this fieldwork, 16 observations were recorded on these sites distributed on the zone (Fig. 3). The time and the difficulty of access prevented us from going on the totality of the working area. Table 1 indicates the metric coordinates of the points founds on the ground.

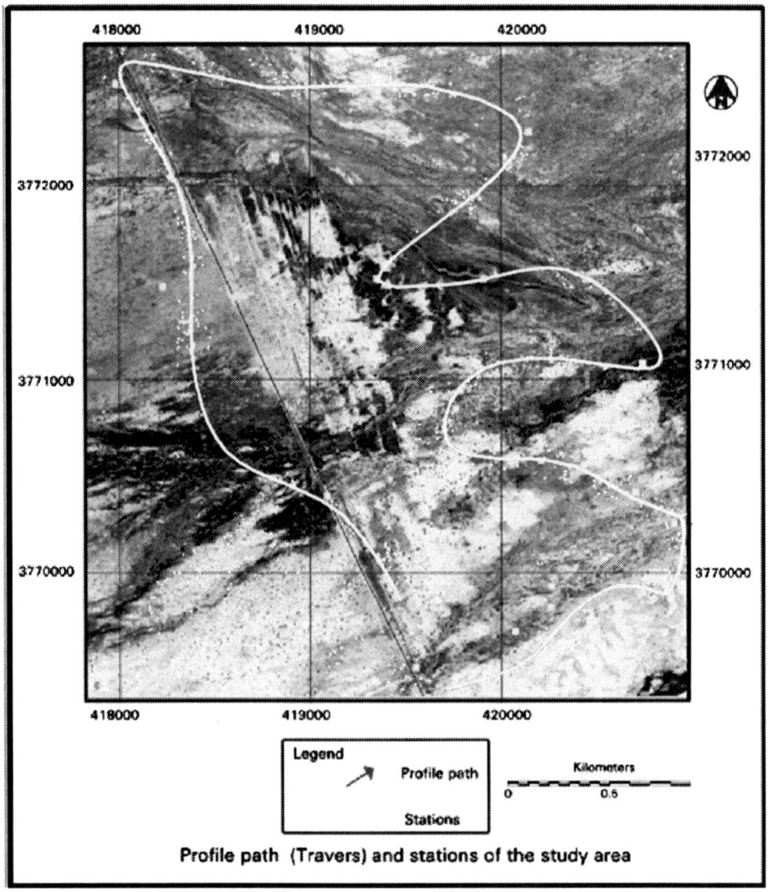

Fig. 3: Profile path and stations of the study area

Table 1: List of stations surveyed and their coordinates

Station	X	Y	N° de photo
Station 1	419483	3769611	P1010061_P1010065
Station 2	420140	3769916	P1010067_P1010069
Station 3	420650	3770034	P1010071_P1010072
Station 4	420742	3770286	
Station 5	420590	3770441	
Station 7	419702	3770686	P1010082_P1010083
Station 8	419652	3770665	P1010084_P1010085
Station 9	420539	3772076	
Station 10	420322	3771472	P1010086_P1010087
Station 11	419837	3771475	P1010088_P1010089
Station 12	419329	3771405	P1010090_P1010098
Station 13	420006	3772151	Photo-067-Photo-070
Station 14	419287	3772393	71-74
Station 15	417629	3773293	78-80
Station 16	418581	3771404	86-87
Station 17	419038	3770459	P10100862_P1010087

3.3 Elaboration of spatial map

To produce a spatial map of the zone of survey, we used the Landsat image as support of interpretation and added the main localities on the satellite data. The digital spatial map was elaborated by using the GIS software (ArcView). The resulting final map indicates the limit of the zone of study with a brown line (Fig. 4).

This spatial map of Oudia area constitutes a good tool of surveillance and operational intervention on the terrain. It permits to direct and orient the teams on the field works and to prepare the plans of action for the struggle against the desertification by taking into account the characteristics of the zone. This work makes a contribution in the field of fight against the desertification by using geographical information.

3.4 Data processing

The different types of image processing and enhancement techniques are carried out by using ERDAS Imagine software to extract, build geo-database for the desertification mapping. The contrast stretching, spatial filtering, an automatic classification and supervised classification, as well as a treatment on the enhancement of radiometric contrast of the image, are used to extract and appear as details information for mapping.

The validity of the training data is evaluated both from visual examination and from quantitative characterization. The spectral signatures of the main features such as sand dune, sabkha and barkhan are evaluated. Therefore, the methods of data analysis and interpretation include unsupervised classification, land cover classification through the highest likelihood method and visual interpretation of different classes of imagery within landscape.

Previously, the team also carried out a visual interpretation of the image under the supervision of the experts who participated in the workshop by using transparent paper.

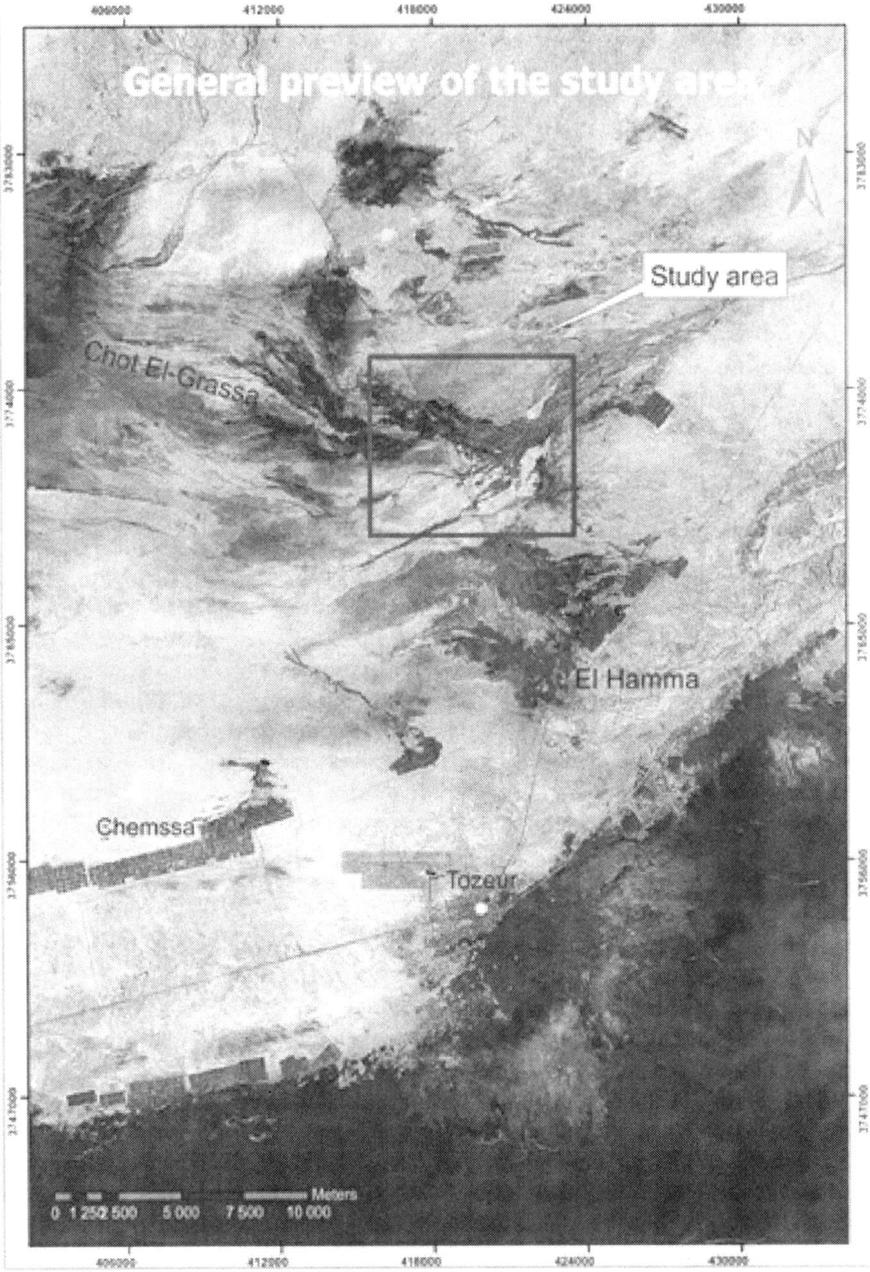

Fig. 4: Spatial map of the study area (Image Landsat)

At the beginning the visual interpretation helped a lot for identifying the different objects present on the ground. During the field work we made a comparison between the reality of ground and the enhanced satellite images.

Figure 5 indicates examples of the type of soils checked on the ground and appears clearly in the classes of land occupation retained during classification.

The dynamic vision of the desertification process is important for the identification of the zones that are already affected, as for those which have a risk in the future. The scale suggested by the team workshop is 1/10,000.

Fig. 5: Results of supervised classification and field verification of Oudia area

4 Results and discussion

4.1 Initial findings

4.1.1 Morphodynamic map

The Morphodynamic map has been established by an approach based on the photo-interpretation satellite images Quick Bird. This interpretation permits to study the units in relation with the soil morphology. Such a detailed geomorphological cartography can contribute to better understanding these dynamics and also better evaluating the techniques of fight against the blocking sand that are applied.

The study area contrasts currently with its environment; this aspect is mainly related to the origin of its soils and due to the acceleration of the wind action.

The Multiple projects of fight against sand encroachment undertaken here also contributed to change the landscape.

We stopped at the El-Maleh Wadi (Fig. 6). It is a wadi with a recurrent flow, 4 to 5 fillings per year with a debit of 18 l/s. This unit is characterized by an ecosystem of contacts between the contributions of river and lake. The flood is going to bring the clay. Black elements in Sebkha, which constitute wet river wash, can be seen. That comes from the laundry of a phosphate layer. When phosphate was washed, the water carried the particles, which are composed about 63% of phospho-gypsum.

These elements were carried by the wind and were deposited on the dunes. A little distribution of nebkhas and rebdou in the zone were observed and recorded but not wide distributed over the area. Stream networks are discriminated on the enhanced satellite images but covered in some places by vegetation.

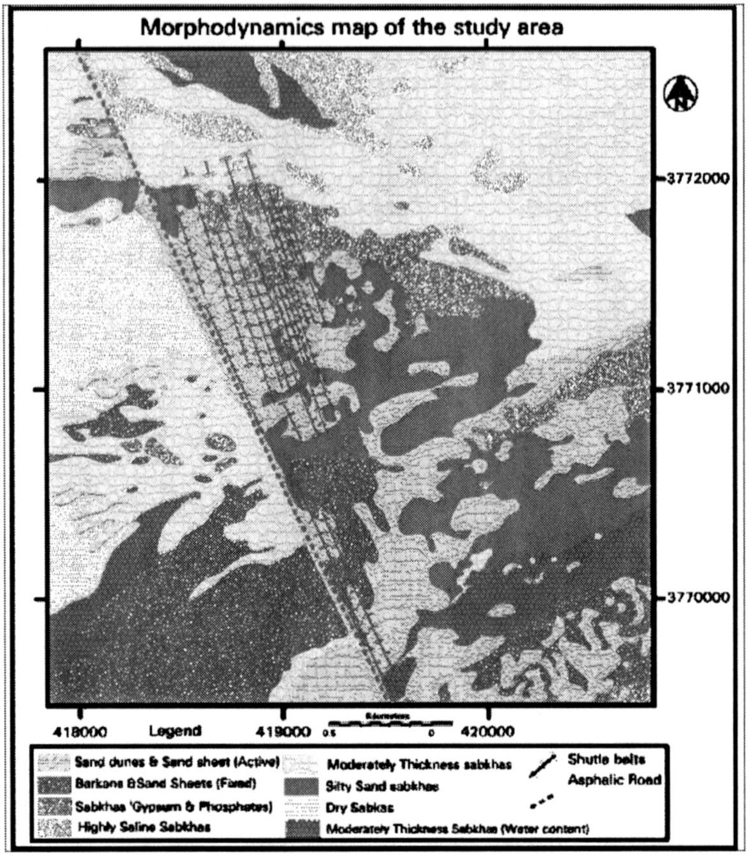

Fig. 6: Morphodynamics map of Oudia area

Globally, in some places the soil is composed of gypsum and sand and the environment is very rich in gypsum. The soil type's distribution in Oudia area, which is investigated in field, is:

1. Halomorphic soils: They characterize the north and south parts of the region of study. They are based on alluvial materials with sand-muddy texture to limono-sandy. Especially during dry season, a form of rash of salts due to the increase of the very salty groundwater and full in gypsum appears. At the end of the dry season, this column of salt and clay disintegrates under the effect of drought. The result is the formation of a superficial horizon of few centimeters in thickness (pseudo-sand). The halomorphic soils show by each place the presence of very varied gypsum accumulations.
2. Gypsum soils: They are soils highly rich in gypsum (rate of 70%), little or not arable and structural only a skinny natural vegetation, serving grazing notably dromedaries.

4.2 Land occupation map

Land cover map is one of key parameters in environmental studies. The study area is distinguished by weak depth and weak content in organic matter. It is dominated by a pseudo steppe with psamophiles and gypsophiles species, characterized by three states (50–200 cm, 15–50 cm and 0–20 cm). This steppe ensures a soil covering of about 20% in the best situations.

The primitive vegetation is a tamarix; but we can generally distinguish three plant groupings, in relation with the type of soil:

- Grouping on the salty formations: Haloconemum strobiloceum, Arthrocnemum indicum, Salsola siberica, Salsola tetrandra, Sueda mollis.
- Grouping on the sandy formations: Stipagrostis pungens, Retama raetam, Calligomum. Aristida pungens is indicating of mobile sand. The presence of this species indicates that sand moves.
- Grouping on the gypseous formations: Astragalus armatus, Zygophyllum album, Erodium glaucophyllum.

The offered pasture is generally poor and constitutes only a summer supplement, in particular for the dromedary. The different classes of soils and their use are observed in Fig. 7. Land use types mapped in this study are: sand dunes, barkhans, sabkhas gypsum and phosphates, highly saline sabkhas, moderately thickness sabkhas, silty sand sabkha, dry sabkha, wet moderately thickness sabkha and rangeland. Mainly, the study area is considered rangeland and grazing for camels.

4.3 Map of vulnerability to the desertification

The conservation of natural resources and the protection of the public infrastructures are the best means of fighting against desertification.

The objective of establishing desertification sensitivity maps is very important to provide support to local decision-makers for better understanding the processes in progress, planning the interventions of attenuation of desertification and monitoring their efficiency.

An integration of remote sensing, GIS and field measurements data is very useful to constrict an establishment of a vulnerability map of Oudia area in this work.

By visual interpretation of enhanced satellite images, desertification units were recognized. Personal knowledge of the same land use, similar physical conditions (climate, soil, vegetation) and common characteristics of land degradation (types and class of desertification) were very important to achieve this work.

Adopted methodologies are based on the analysis of the dynamics of the landscape. Figure 8 represented the result of images processing that shows the vulnerability map of desertification.

The wind represents the primary mechanical agent of desertification. It was the cause for which sand deposits became dunes. In the process of elaboration of desertification sensitivity map, the sand chart constitutes an important element.

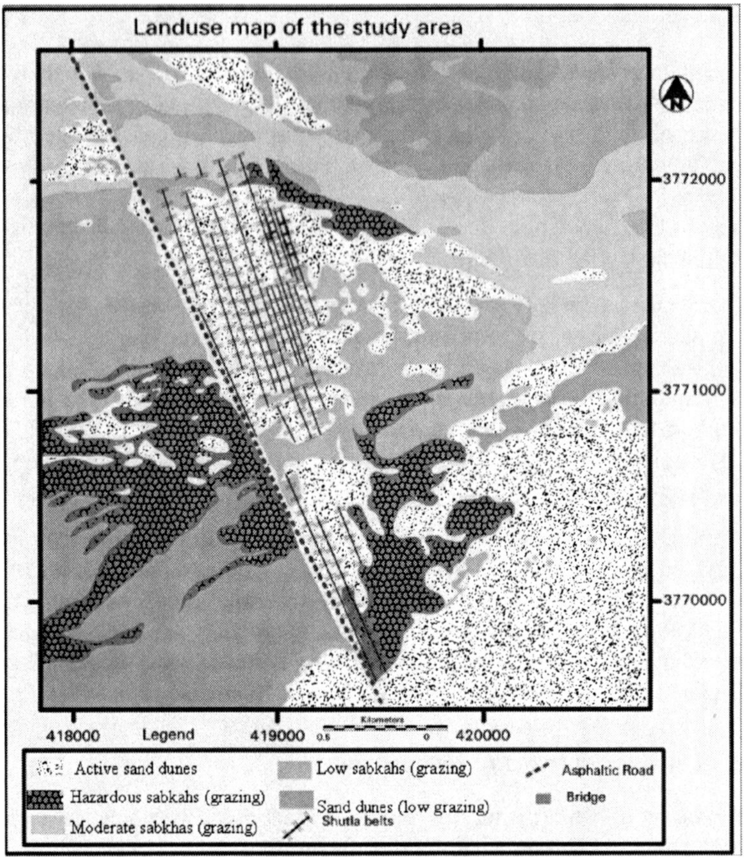

Fig. 7: The land use/land cover map of Oudia area

This map elaborated from Quick Bird data shows the big types of dunes, which are discriminated in step with their reflectance:

The final result was an image map of vulnerability of desertification, which gives good indication of areas affected by desertification and those under risk of desertification. The areas were divided into seven classes including high, moderate and low hazardous.

Active sand dunes
Moderate sand dunes
Low sand dunes
Highly hazardous sabkhas
Moderately hazardous sabkhas
Low hazardous sabkha
Flash flood

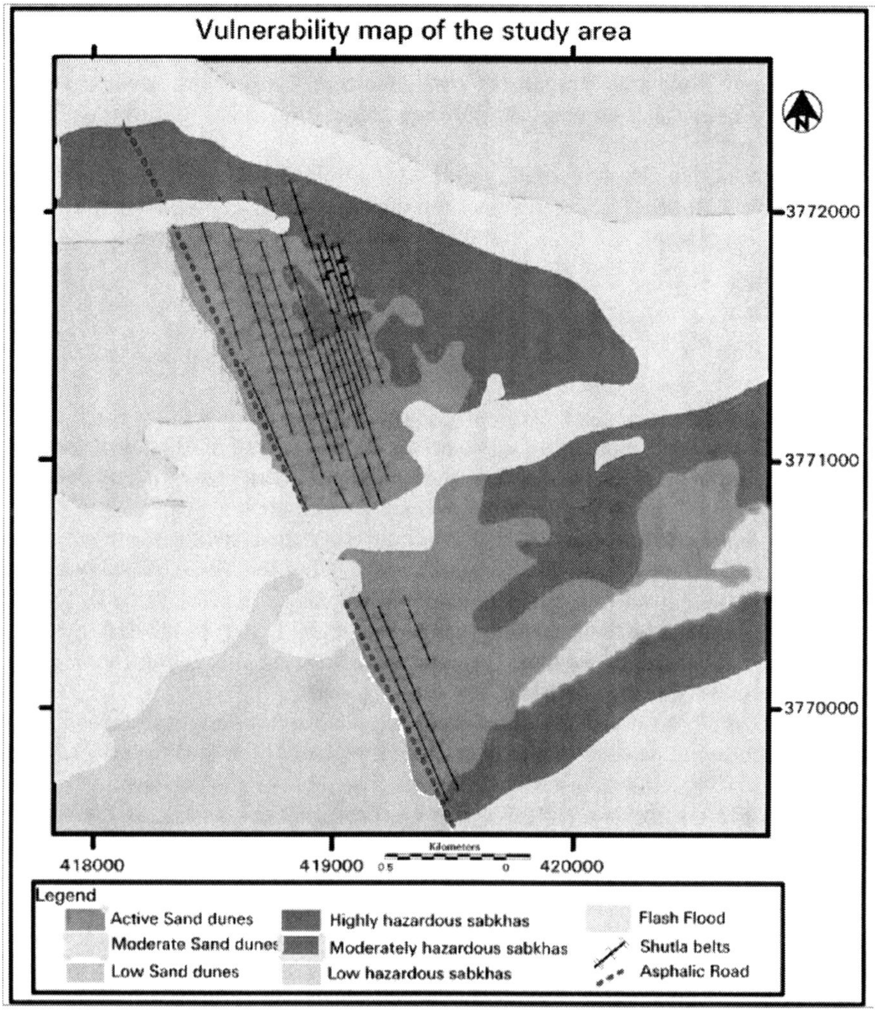

Fig. 8: The sand cover map of the study area

On the ground, we saw fences and lineaments, which are used for the fight against the desertification. But the barriers were built in perpendicular with the dominant wind direction. In this zone, there are two dominant sectors of the wind: the West during the winter and the East during summer. However, 65% of wind comes from the east.

It appears that the fences amplify the trapping of a lot of sand. We implanted 23 parallel fences on 4 km with 50 m of interval between two fences (Fig. 7). Since the beginning we made seven generations of fences. It is the dynamic due to the human action of sand trapping; people think that it is negative. The question is if these fences embarrass the existing plantation.

The zone between two fences is investigated and verifies if the sand is going to become a plateau of sand. In our opinion, it is necessary to define the situation before to know if the intervention was justified or not. To look for the alternative, it is necessary to make a comparison between the natural sector and the anthropological sector.

We remind that the vulnerability map to the desertification remains temporary and that it will be necessary to conduct a more detailed survey in the future, while using data, for example, on soils, climate, soils sensitivity to the erosion, vegetation maps.

5 Conclusion

This paper outlines a number of pilot applications of remote sensing for mapping desertification in arid areas. The applications of the remote sensing and the geographical information systems constitute good means of investigation to help decision-makers in the fight against the desertification and the reasoned management of natural resources. A reliable information and constantly update constitutes a need and a guarantee for a sustainable development of natural resources. These techniques allow us measuring the impact of the phenomenon on the people, as well as on the natural environment. As regards the elaboration of the Vulnerability map to the desertification we insist on the need of harmonizing and standardizing the approaches and the methods used everywhere in Africa.

It is necessary to revise the current models concerning the physical, ecological and socioeconomic aspects of the degradation of lands, for their integration in a framework of dynamic model and instruments of support for the decision.

The visual interpretation combined with the digital processing of the satellite data with high-resolution Quick bird and Landsat permitted to classify the territory studied in its various components at the geomorphological level.

This study shows that the use of the satellite data with high resolution allows the elaboration of a cartography that records the state of degradation of the ground under the influence of the desertification. These documents are essential for the rehabilitation of the sector of Oudia, today strongly threatened by the desertification.

Acknowledgements

The authors thank the organizers of this international Workshop, Prof. Marini Alberto of the laboratory of TELEGIS at the University of Cagliari and Prof. Mohamed Talbi of the Institute of the Arid Areas of Medenine (WILL GO), for their invitation and the good organization. They also thank Mr. Nabil Gasmi for his help on the fieldwork. Thanks to all the team that directed us during the workshop, in particular Prof. Ben Azouz, Prof. Ashour, Prof. Ozer, Prof. Abdellaoui and Prof. Kevin White. We would also like to thank the colleagues who shared with us this pleasant stay in the beautiful Tozeur. A special thank to the mayor of Tozeur and the employees of the Hotel and restaurant and museum of Dar Chreait for the quality of service.

Bibliography

Blaikie P. and Brookfield H. 1987. *Land degradation and society*, Longman, London.

Clarke and Noin 1998. In: J. Clarke and D. Noin, Editors, *Population and environment in arid regions*, UNESCO/Partenon Publishing Group, Paris, pp. 1–18.

Global Land Cover Facility: ftp://ftp.glcf.umiacs.umd.edu/glcf/Mosaic_Landsat/N-32/N-32-30.ETM-EarthSat-MrSID/

Laboratorio TELEGIS, Dipartimento Scienze della Terra, Università di Cagliari - http://www.unica.it/telegis/

Murray S., Burke L., Tunstall D., and Gilruth, P. 1999. *Drylands population assessment II. UNSO.*

Strahler A. and Strahler A. 1992. *Modern physical geography*, Wiley, New York.

Talbi M., Gasmi N., and Marini A. 2006. *Présentation générale de la région de Tozeur (Le Jerid), Cartographie des paysages arides pour la lutte contre la désertification, Tozeur, Tunisie*. 9–19 décembre 2006, p. 28.

UNCCD. 2004a. FAQ. http://www.unccd.int/knowledge/faq.php#answer2.

UNCCD. 2004b. FAQ. http://www.populationdata.net/images/cartes/tunisie.jpg

United Nations. 1994. *Elaboration of an international convention to combat desertification in countries experiencing serious drought and/or desertification, particularly in Africa.* Final text of the Convention. www.unccd.int/convention/text/pdf/conv-eng.pdf.

SAND DUNES EVOLUTION AND DESERTIFICATION IN SOUTHEASTERN MOROCCO: A NEW APPROACH TO AN OLD PROBLEM

AMINA MALAKI, MOHAMED EL WARTITI, AND ABDELILAH EL GHANNOUCHI
Département des Sciences de la Terre, Faculté des Sciences UFR R.M.R.E.E. Rabat
Université Mohammed V –Agdal-Rabat
Amina_ok@yahoo.fr

Abstract

This paper presents an analysis of natural and anthropogenic factors responsible for increasing sanding up and desertification phenomena in southeastern Morocco. A new relationship between humans and the environment has developed through the introduction of modern lifestyles in oasis areas, accompanied by increased road systems, irrigation canals, motorization, dams, etc. This has led to a generally unbalanced situation. The main environmental changes that have occurred, such as sanding up and desertification, are therefore more due to human pressure and accompanying technological modifications, along with resource mismanagement, than to unfavorable climatic conditions. Human activities are therefore mainly responsible for the environmental imbalance that has occurred in southeastern Morocco in recent decades.

Keywords: Anthropogenic factors, sanding up, desertification, the environmental imbalance, Southeastern Morocco

1 Introduction

Desertification is considered one of the most critically important issues in many countries (Rodriguez, 1997). The United Nations Convention to Combat Desertification (UNCCD) defines it as "land degradation in arid, semi-arid areas resulting from various factors, including climatic variations and human activities" (http://www.unccd.int). Particularly, in West Africa, desertification is a problem of major importance that creates a situation of misery together with drought and poverty. Indeed, these regions are now suffering startling land degradation. Following the reduction of the vegetation cover, many dunes which were formerly stabilized by vegetation, are now reactivated. Currently, on the edge of the Sahara desert, moving dunes threaten the existence of many farmlands, water points, settlements and transport links.

A. Marini and M. Talbi (eds.), *Desertification and Risk Analysis Using High and Medium Resolution Satellite Data*,

In order to fight against sand dunes encroachments and desertification in an efficient manner and ensure successful and sustainable projects, a good understanding of the driving mechanisms through an integrated approach is required, including the mapping and monitoring of sensitive zones (Bodart, 2004).

Although some desertification is attributable to natural causes (long term climate change, droughts, and localized weather events), most instances can be attributed to human activity. Drylands are characterized by limited availability of arable land, limited and highly variable rainfall and scarcity of water resources. The demands placed on land and water resources by rapidly expanding populations, through agricultural intensification, urbanization and industrialization have combined to intensively exploit these natural resources. The principal direct causes of degradation in drylands are:

Removal of vegetation through cutting and uprooting trees and shrubs, plowing previously uncultivated and marginal land for annual cropping, poor soil management practices following harvest, overgrazing natural rangelands (Fikri et al., 2004).

Intensification of cultivation using inappropriate cultural practices that degrade soil fertility and encourage erosion and through overuse of irrigation and poor drainage, leading to rising water-tables, water logging and salinization.

Increasing pressure on natural resources and hard climatic conditions generate terrestrial ecosystem dysfunctions. These effects are amplified by an inappropriate use of the existing natural resources, thereby leading to regression of forest clumps, decrease of water supplies and water pollution, soil degradation, and hence to desertification and reduction of animal and plant diversity. Such disruptions affecting the biological resources and the earth supplies also affect the lives of the people who end up leaving their fields to move to neighboring cities or emigrate. New techniques of environmental remediation are available to enhance production and protect ecosystems, but unfortunately, social and economic issues are reducing these advances to nothing. Arable land expansion leads to deforestation and to overgrazing, which causes water and wind erosions. All these factors along with drought, topography, soil and vegetation overuse, as well as farming systems, contribute to the desertification process in semiarid zones. The chemical, physical and biological properties of the soil underline important changes, which play an essential role in causing vulnerability to desertification due to human activities. This phenomenon prevents the reconstitution of the vegetal cover and can be considered as the ultimate form of soil degradation (Fikri et al., 2004).

Radar, remote sensing and interferometers have proved their utility in a wide range of earth observation studies in particular for the study of aeolian transport processes and detection of active sand dunes (Bodart, 2004).

2 Sand dunes evolution and desertification in Draa valley and Hannabou sector

Preliminary results of two case studies are presented. The two areas Draa valley and Hannabou sector) are located in southeastern Morocco on the northern border of the Sahara desert (Desjardins et al., 1999).

2.1 Draa valley

The Draa valley takes shape, downstream from the city of Ouarzazate, at the junction of the Dades and Ouarzazate wadies (Fig. 1).

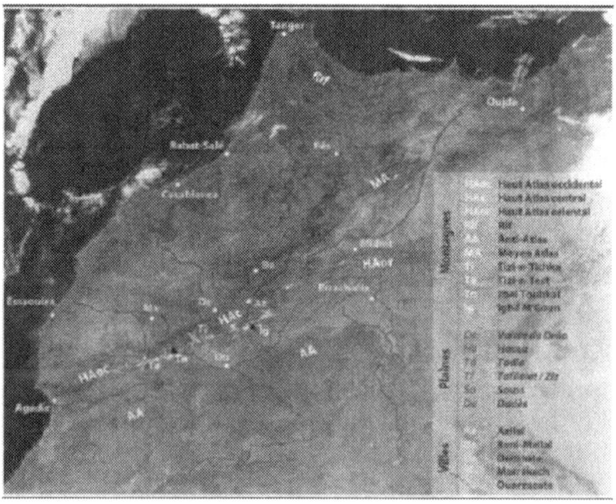

Fig. 1: Draa valley on the Satellite photo of 25.01.2002 by the sensor MODIS (NASA)

In its middle portion, i.e. from Agdz to M'Hamid, the presence of Ordovician shale allowed the development of wide depressions surrounded by sandstone ridges.

These ridges protect six palm groves separated by transverse valleys, named "cluses" or "foums". In this region, rainfalls present a North-South gradient decreasing from 300 to 40 mm per year from upstream to downstream and show strong variations from year to year. In addition to this lack of water, the vegetation is threatened by anthropogenic degradations, such as trampling and overgrazing. These phenomena led to sand dunes encroachment, particularly around the palm groves (Benmohammadi et al., 1998).

Figure 2 shows the RGB color composite resulting from the tandem pair of 18–19 May 1996 as well as small sections of an IKONOS image and field observations, allowing validating the interpretations. The zones with coherence losses were systematically correlated with active dunes or, in general, with processes of sand dune encroachment. In addition, specific areas as villages, oases or roads were identified in order to help fighting desertification in Morocco (Benmohammadi, 1995).

2.2 Hannabou sector, Tafilalet

Progress of sanding over for some palm groves in the southeastern part of Morocco has been studied in 1999 with the help of 2 SPOT imageries acquired in 1986 and 1995. Recently, images from military satellites (acquired between 1963 and 1980) and ASTER imagery (2003) enriched our data, thus we can now analyze the phenomena over a period of 40 years (Desjardins et al., 2005). The spatial resolution

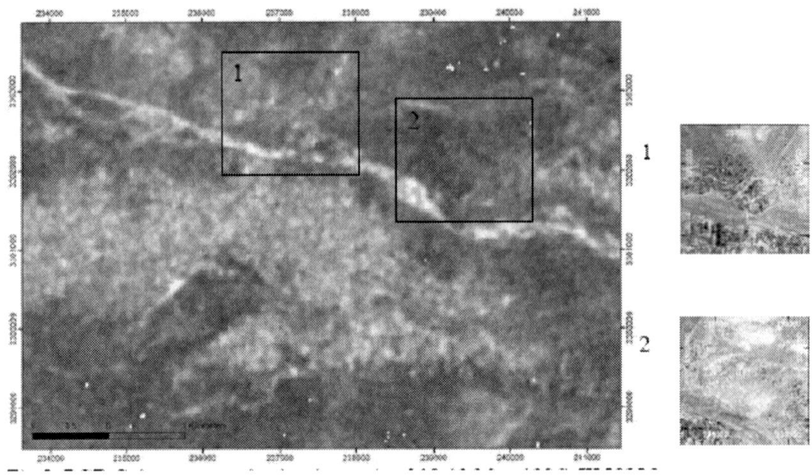

Fig. 2: RGB color composite (tandem pair of 18–19 May 1996) IKONOS image (29 October 2001) – field observations (November 2003), 1: M'Hamid village threatened by sand dune encroachment, 2: active field dune in the east of M'Hamid

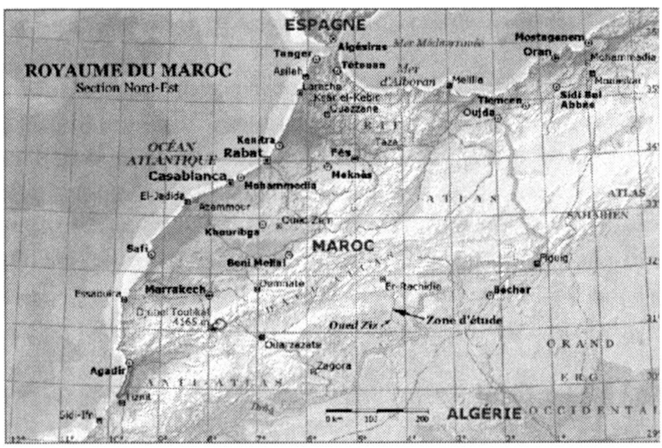

Fig. 3: Localization of the sector studied among the natural areas and the principal agglomerations of Morocco

of these imageries is fine enough, not only for measuring orientation and displacement of dunes, but also for better understanding the dynamic of sanding (Benalla, 2003) over for the studied area (Hannabou sector, the western part of Tafilalet: Figs. 3–5).

Digital image processing, combined with good spatial resolution (Desjardins et al., 2005) permitted to have more robust measures concerning these orientation and displacements (Fig. 7). In fact, winds coming mostly from the SW (locally named Saheli wind) generated an average displacement of about 15 m/year in the NE direction.

Furthermore, this study helped us emphasizing the local origin of sands in forming dunes in this area (Fig. 6).

Fig. 4: Three-dimensional sight towards north of the area. The SW part corresponds to the edge of the Anti-Atlas strongly eroded. The NE part represents a plane surface of Tafilalet. The red illustrates the plant environment of palm groves, broadsides in the east by the valley of the Rhéris wadi, while, in the NE corner, we can see the outlet of the Ziz wadi in the plain of Tafilalet

Fig. 5: Stranding of the palm plantation of Hannabou

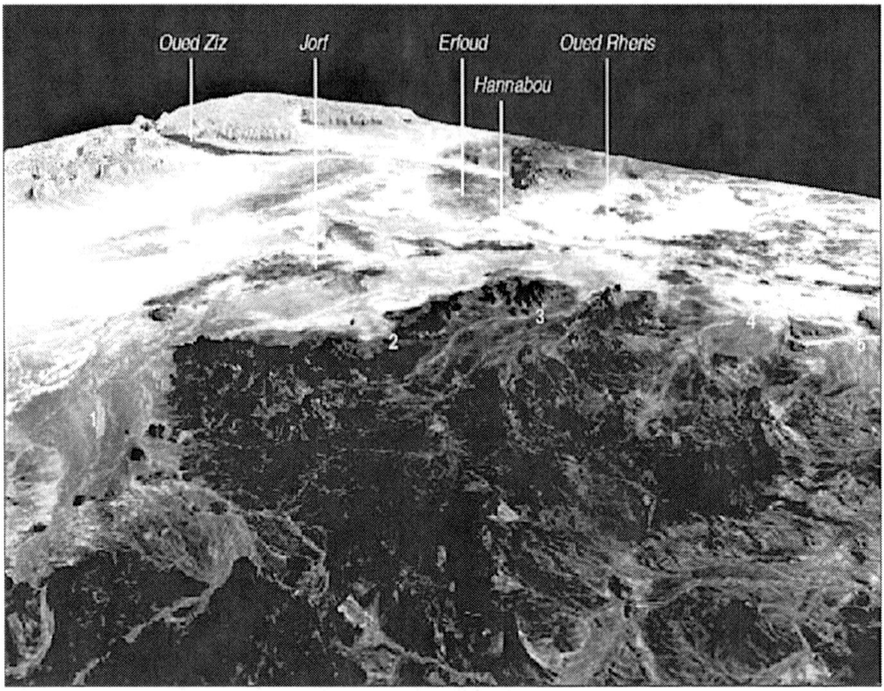

Fig. 6: Sand supply of local origin. The NE orientation of this three-dimensional vision of the landscape makes it possible to identify the sources of the sands brought by the local wadis

Fig. 7: Orientation and displacement of barkhanes immediately to the SW of the palm grove of Hannabou, from 1964 till 2003. The full lines of different colors, represent for each studied period (1970–1977, and since 1980) the distance covered by certain dunes according to the prevalent winds: the sahéli SW-NE and the chergui NE-SW. For the other periods, when the wind shaves different directions over shorter periods (2 to 4 years), displacements of the dunes, much more limited, are represented by hyphens

3 Conclusion

Newly developed tools such as remote sensing, GIS, and computer-assisted expert systems can contribute to the planning and implementation of concrete measures to combat desertification that also greatly improve dryland productivity. Capitalizing on the positive aspects of sunshine and temperature of dryland environments, the provision of water for irrigation allows the economic production of crops and other agricultural products on lands reclaimed from the desert areas. Reclamation often requires heavy investments in earth moving and irrigation infrastructure construction.

In other areas, non-conventional sources are needed. Desalinizing seawater using innovative energy sources such as solar power, tidal movement, or even bio-gas offers interesting opportunities for development in areas without renewable sources of surface or groundwater.

Since heat, moisture stress, and salinity are the major abiotic constraints to vegetative growth in arid and semi-arid environments, enhancing the ability of introduced plant species to withstand these constraints is a major concern in reclamation and rehabilitation efforts. Genetic engineering and other biotechnology tools offer considerable promises, together with improved conventional plant breeding methods. At the same time, there is need to develop appropriate land and water management techniques to prevent erosion, fertility loss and salinization so that the production system becomes sustainable.

Desertification is a reversible process, but action must be taken immediately to reverse the process in areas where the threat is greatest before the process reaches its conclusion and there is no longer the chance of recovery. In the arid and semi-arid zone has the highest priority because it is here that removal of natural vegetation and inappropriate cultivation methods are degrading and depleting valuable and limited biological, soil, and water resources at the fastest rate.

Bibliography

Benalla M. 2003. *Étude morpho-dynamique de l'évolution des dunes du Tafilalet (Maroc): apports de la sédimentologie et de l'imagerie aérienne et spatiale.* Thèse de doctorat, Université Mohammed V-AGDAL, faculté des Sciences de Rabat.

Benmohammadi A., Benmohammadi L., Morel A., Chardon M. and El Hanbali M. 1998. *Ensablement et désertification dans la vallée moyenne de l'oued Drâa: le mythe de la fluctuation de la limite nord du désert.* Africa Geoscience Review, n.**5**, 11–22.

Benmohammadi L. 1995. *Désertification et ensablement dans la vallée moyenne de Drâa: étude géomorpholgique des formations dunaires (Sud marocain).* Thèse doctorat, Grenoble, Université Joseph-Fourier.

Bodart C. 2004. *Etude de l'ensablement dans la région de Gouré (sud-est du Niger): processus et apport de la télédétection.* Master Dissertation, Faculté des Sciences, Université de Liège.

Desjardins R., Alem E.M., Rognon P., Maudin P., Ndiaye I. and Benalla M. 1999. *Sur l'ensablement des palmeraies. Apport de la télédétection numérique pour un secteur menacé dans le sud-est du Maroc.* Actes des huitièmes journées scientifiques du « Réseau Télédétection » de l'Agence universitaire de la Francophonie, Lausanne, 22–25 novembre 1999, 127–139.

Desjardins R., Rognon P., Benalla M. and Alem E.M. 2005. *Progrès importants des études sur l'évolution des dunes grâce à l'utilisation combinée des satellites civils et militaires: l'exemple du Tafilalet (Maroc),* in Science et changements planétaires/Sécheresse, n.**16** (3), 153–164.

Fikri Benbrahim K., Ismaili M., Fikri Benbrahim S. and Tribak A. 2004. *Problèmes de dégradation de l'environnement par la désertification et la déforestation: impact du phénomène au Maroc*, in Science et changements planétaires/Sécheresse, n.**15** (4), 307–320.

Gassani J. 2004. *Apport de la télédétection à l'analyse et à la gestion des risques naturels sur la zone du lac d'Aleg (Brakna, Mauritanie).* Master Dissertation, Faculté des Sciences, Université de Liège.

Ohammadi A., Benmohammadi L., Morel A., Chardon M. and El Hanbali M. 1998. *Ensablement et désertification dans la vallée moyenne de l'oued Drâa: le mythe de la fluctuation de la limite nord du désert.* Africa Geoscience Review, n.**5**, 11–22.

Ozer A., Legat A., Schyns J.C., Petit S., Salmon M., Goossens R., De Dapper M. and Ost L. 2002. *Etude par télédétection de l'ensablement dans la province de Ouarzazate, Maroc,* Project report, Liège.

Ozer P. 2000. *Les lithométéores en région sahélienne*, GEO-ECO-TROP, 24, 1–4.

Reynolds J.F. and Stafford Smith D.M. 2002. *The Dahlem desertification paradigm: a new approach to an old problem.* In: *Global Desertification: Do human cause deserts?* Edited by. J.F. Reynolds and M. Stafford Smith. Dalhem University Press, Berlin, Germany, 1–21.

Rodriguez A. 1997. *Rural Poverty and Natural Resources in the Dry Areas: The Context of ICARDA's Research.* Working Paper. Aleppo, ICARDA.

THEMATIC MAPPING USING QUICKBIRD MULTISPECTRAL IMAGERY IN OUNG EL-JEMEL AREA, TOZEUR (SW TUNISIA)

FRANCESCO G. DESSÌ[1] AND ABDOUL JELIL NIANG[2]
[1]*Earth Science Department, TeleGis Lab, Via Trentino 51, 09127 Cagliari, Italy*
fdessi@unica.it
[2]*Laboratoire de Géomorphologie et Télédétection, Allées du 6 Aout 2, B11-4000 Liège, Belgium*
ajeniang@yahoo.fr

Abstract

Standard ground methods of land use mapping are labor intensive, time consuming and are done relatively rarely. In recent years, satellite remote sensing techniques have been developed, with increasing availability of very high resolution satellite imagery, which have proved to be of immense importance for preparing accurate land use and land cover maps. In case of inaccessible region, this technique is perhaps the only method of obtaining the necessary data on a cost and time-effective basis.

The aim of this work was to investigate the use of Quick Bird imagery for thematic mapping in Oung El-Jemel area, close to the town of Tozeur (SW Tunisia), supported by field verifications data.

Keywords: Thematic mapping, Quick Bird, NDVI

1 Introduction

The Oung El-Jemel area is located in SW Tunisia (Fig. 1) and takes its name from a unique rock formation near the salt lake Chot el Gharsa. This area has been famous among locals for its camel-like shape and recently has become a tourist attraction following a number of movies being filmed there.

The principal process is aeolian, with a variety of mobile sand dunes and semi-fixed sand dunes. The moving sand is also forming erosion landforms, such as yardangs. In addition to these aeolian processes, fluvial erosion is forming small gullies along the shoreline of the Chott el Rhasa, which consists of exposures of Villafranchian sediments.

2 Quick Bird satellite data

Quick Bird offers panchromatic and multispectral imagery with the highest spatial resolution currently available within the satellite sensors. The Quick Bird satellite has panchromatic and multispectral sensors with resolutions at 61–72 cm and 2.44–2.88 m,

A. Marini and M. Talbi (eds.), *Desertification and Risk Analysis Using High and Medium Resolution Satellite Data*,

Fig. 1: Study area and "camel" rock

respectively, depending upon the off-nadir viewing angle (0–25°). The sensor has coverage of 16.5e19 km in the across-track direction. In addition, the along-track and across-track capabilities provide a good stereo geometry and a high revisit frequency of 1–3.5 days depending on latitude. Quick Bird technical notes are available on line at the web site http://www.digitalglobe.com/product/product_docs.shtml.

The panchromatic sensor collects information at the visible and near infrared wavelengths and has a bandwidth of 450–900 nm. The multispectral sensor acquires data in four spectral bands from blue to near infrared (NIR). Both panchromatic and multispectral sensors offer 11 bit (2,048 grey levels) resolution. The Quick Bird imagery products are available at different processing levels (basic, standard, and ortho) serving the needs of different users.

In this work we have used a Quick Bird image acquired under clear sky conditions on 18th February 2004.

3 Methodology

3.1 Satellite data processing

The idea that lies behind our analysis is that features could be enhanced and efficiently studied by exploiting the high spatial resolution of satellite Quick Bird panchromatic data and the multispectral properties of the four spectral channels. The methodological approach (shown in Fig. 2) adopted for the enhancement and extraction of homogenous areas is mainly based on photointerpratation (Richardson and Wiegand, 1977; Myneni et al., 1995).

Data fusion refers the process of combining multiple images of a scene to obtain a single composite image. The different images to be fused can come from different sensors of the same basic type or they may come from different types of sensors. The composite image should contain a description of the scene more useful than those provided by any of the individual source images. In the current cases under investigations the Quick Bird panchromatic and multispectral images were fused (Zhang, 2004) by using pan-sharpening in ENVI® software.

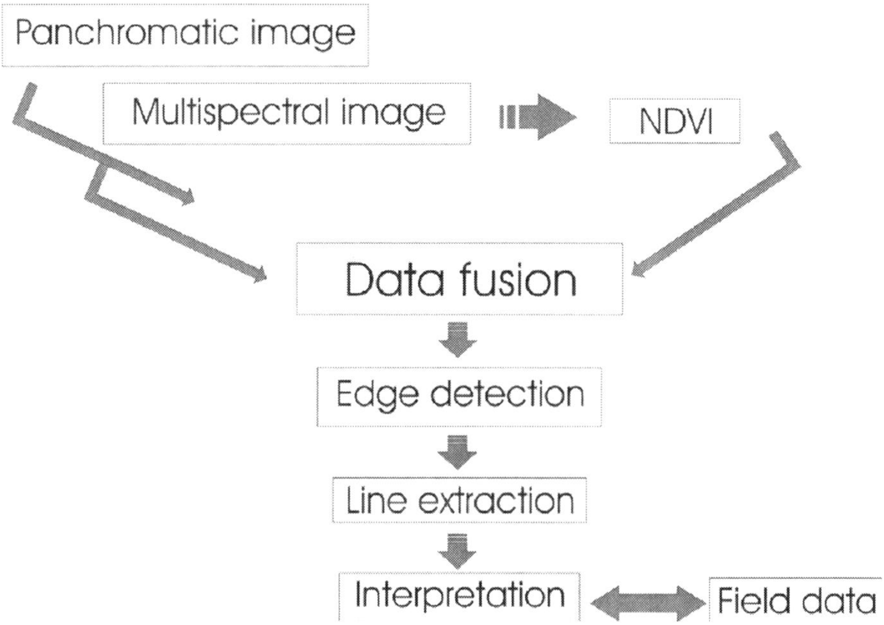

Fig. 2: Flow chart of the methodology adopted

3.2 NDVI

Additionally, on the basis of remotely sensed data, vegetation be suitably identified by exploiting vegetation indexes (Bannari et al., 1995; Elvidge and Chen, 1995) that are spectral combinations of different bands. Such indexes are quantitative measures, based on vegetation spectral properties, that attempt to measure biomass or vegetative vigour. Vegetation indexes are mainly derived from reflectance data from the Red and near infrared (NIR) bands. They operate by contrasting intense chlorophyll pigment absorption in the red against the high reflectance of leaf mesophyll in the near infrared. The simplest form of vegetation index is the ratio between two digital values from the red and near infrared spectral bands.

The most widely known and used ratio-based index is the normalized difference vegetation index (NDVI) (Rouse et al., 1974; Carlson and Ripley, 1997). The NDVI is calculated as the ratio of the difference between the near infrared and the red band and the sum of the two bands by using the following formula:

$$NDVI = (NIR - Red) / (NIR + Red)$$

The normalization of the NDVI reduces the effects of variations caused by atmospheric contaminations. NDVI is indicative of plant photosynthetic activity and has been discovered to be related to the green leaf area index and the fraction of photo-synthetically active radiation absorbed by vegetation. Using this index, the difference between vegetation and non-vegetation is emphasized. Reflectance values for vegetation have their maximum in the near infrared and a minimum in the red spectral domain. High values of the NDVI indicate vegetation and values around 0 non-vegetated land areas (Figs. 3, 4 and 5).

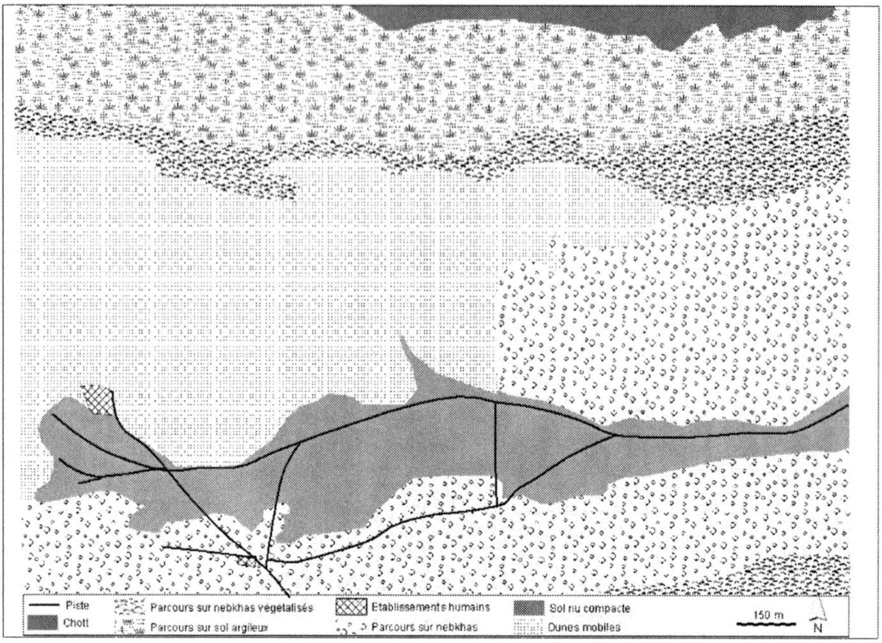

Fig. 3: Homogeneous domains in the Oung El-Jemel area

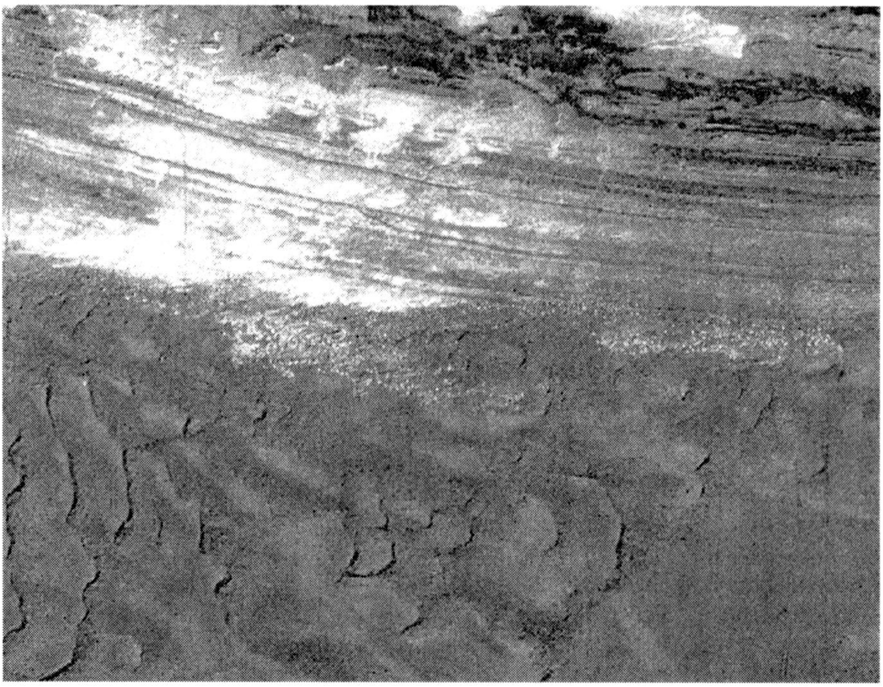

Fig. 4: NDVI for a portion of the Quick Bird image of the study site. Bright values indicate high level of vegetation and dark values non-vegetated areas

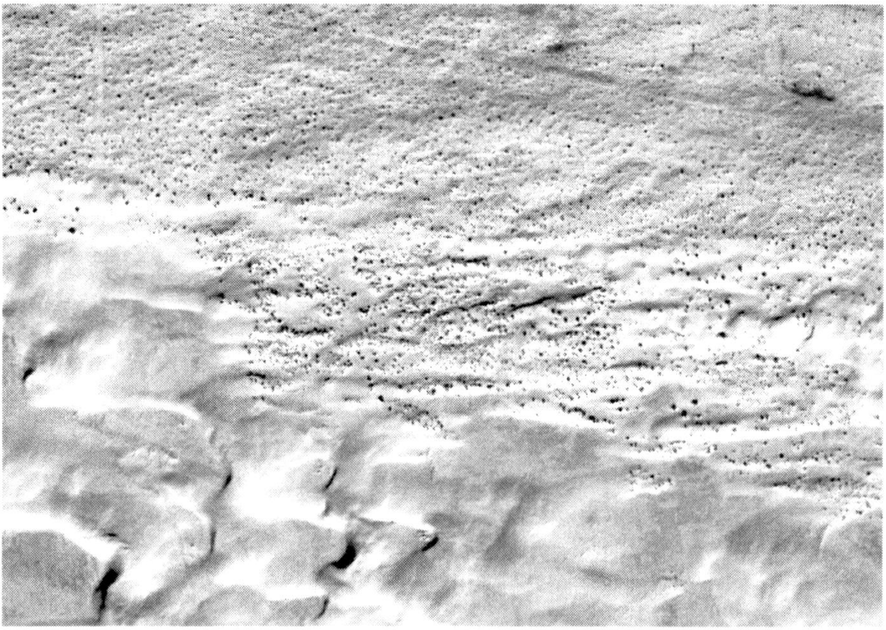

Fig. 5: Portion of Quick Bird scene of the study area. In the southern zone some mobile sand dunes. In the northern area some semi-fixed dunes and sporadic vegetation. FCC 432 with Pan-sharpening

3.2.1 Chott

It is the border of a sebkha covered with halophilic vegetation; a sebkha being an ephemeral salt water body.

3.2.2 Sebkha

A geologic feature, in North Africa, which is a smooth, flat, plain usually high in salt; after a rain the plain may become a marsh or a shallow lake until the water evaporates.

3.2.3 Clayey soil rangeland

Clayey soil rangelands extend along the border of the Chott and are covered with halophilic vegetation. Salty surface crusts are frequently present in this area.

3.2.4 Rangelands on vegetated Nebkhas

Correspond to rangeland on semi-mobile dunes. The vegetation consists of small (<1 m) bushes that allow V shaped Nebkhas development.

3.2.5 Human settlements

Mainly consist of camps of nomads. Among the few building of the area, we can mention the installations constructed for the movie “Stars Wars”.

3.2.6 Nebkhas

Rangelands similar to those on vegetated Nebkhas, but with less vegetation and a higher aeolian activity.

3.2.7 Compact bare soils

Flat area free from sand and with a very small vegetation (Drake, 1997) cover located between the nebkhas and the area of migrating dunes.

3.2.8 Migrating dune

Migrating (or shifting) dunes cover all the area. These barchans extend up to 100 m in width and 10 m in height. This area is free from vegetation cover.

4 Conclusions

Satellite Quick Bird imagery were used in order to assess their capability to detect objects in the study area. The use of data fusion and photointerpretation procedures improves the identification of homogeneous domains.

The integration of results obtained from panchromatic and image fusion products provides valuable information for a detailed physical and geometrical characterisation of the region.

The use of NDVI allowed bettering enhancing surfaces covered by green and herbaceous plants.

Bibliography

Bannari A., Morin D., Bonn F., Huete A.R. 1995. *A review of vegetation indices*, in Remote Sens. Rev. **13**, 95–120.

Carlson T.N., Ripley D.A. 1997. *On the relation between NDVI, fractional vegetation cover, and leaf area index*, in Remote Sens. Environ. **62**, 241–252.

Drake D.A. 1997. *Recent aeolian origin of superficial gypsum crusts in Southern Tunisia: geomorphological, archaeological and remote sensing evidence*, in Earth Surf. Proc. Land. **22**, 641–656.

Elvidge C.D., Chen Z. 1995. *Comparison of broadband and narrow-band red and near infrared vegetation indices*, in Remote Sens. Environ. **54**, 38–48.

Myneni R.B., Maggion S., Iaquinta J., Privette J.L., Gobron N., Pinty B., Verstraete M.M., Kimes D.S., Williams D.L. 1995. *Optical remote sensing of vegetation: modelling, caveats and algorithms*, in Remote Sens. Environ. **51**, 169–188.

Richardson A.J., Wiegand C.L. 1977. *Distinguishing vegetation from soil background information*, in Photogramm. Eng. Rem. Sens. **43**, 1541–1552.

Rouse J.W., Haas Jr., R.H., Schell J.A., Deering D.W. 1974. *Monitoring Vegetation Systems in the Great Plains with ERTS*, in NASA SP-351, 3rd ERTS-1 Symposium, Washington, DC, pp. 309–317.

Zhang Y. 2004. *Understanding image fusion*, in Photogramm. Eng. Rem. Sens., **70** (6), 657–661.

USE OF REMOTE SENSING AND GIS FOR MONITORING AGRO-ECOLOGICAL SYSTEM IN TUNISIAN ARID REGIONS (*STUDY CASE OF MENZEL HABIB*)

HANEN DHAOU[1], A. BELGHITH[2], AND MOHAMED OUESSAR[1]
[1]*Institut des Régions Arides 4119, Medenine, Tunisia*
hanen_dhaou@yahoo.fr
Ouessar.Mohamed@ira.rnrt.tn
[2]*National Centre for Remote Sensing, Tunisia*

Abstract

Since the last decades, desertification has become the major environmental problem in pre-Saharan Tunisia and it is continuously increasing. Climatic aridity and the over exploitation of natural resources in these arid regions are the main causes of rapid degradation of the environment: 25% of the arid and semi-arid regions is highly affected by this phenomena (Ouled Belgacem and Neffati, 1996).

Within the limitations of the field tools, remote sensing offers nowadays new possibilities for characterization and monitoring of the land use and their vulnerability to desertification within different scales.

The objectives of our study are the following:

1. Development of a specific methodology for monitoring the degradation of agro-ecological systems in pre-Saharan Tunisia through remote sensed data.
2. Elaboration of the agro-ecological zoning for the annual evaluation of the change in land use, the state and the dynamic of vegetation.

Keywords: Ardity, drought., Tunisia, ecological zoning

1 Introduction

The study area is situated between the two parallels 34° 02′ and 34° 21′ North and the meridians 9° 05′ and 9° 58′ East, in the observatory of Menzel Habib: ROSEL/OSS.

The global area is about 1,678 km^2. This region is characterized by scarce water resources and diversified but poor fertile soils with a texture vulnerable to erosion.

A. Marini and M. Talbi (eds.), *Desertification and Risk Analysis Using High and Medium Resolution Satellite Data,*

2 Methodological approach

To elaborate an agro-ecological zoning in the region of Menzel Habib, we used a methodological approach based essentially on simplified cartographic intersections (Fig. 1).

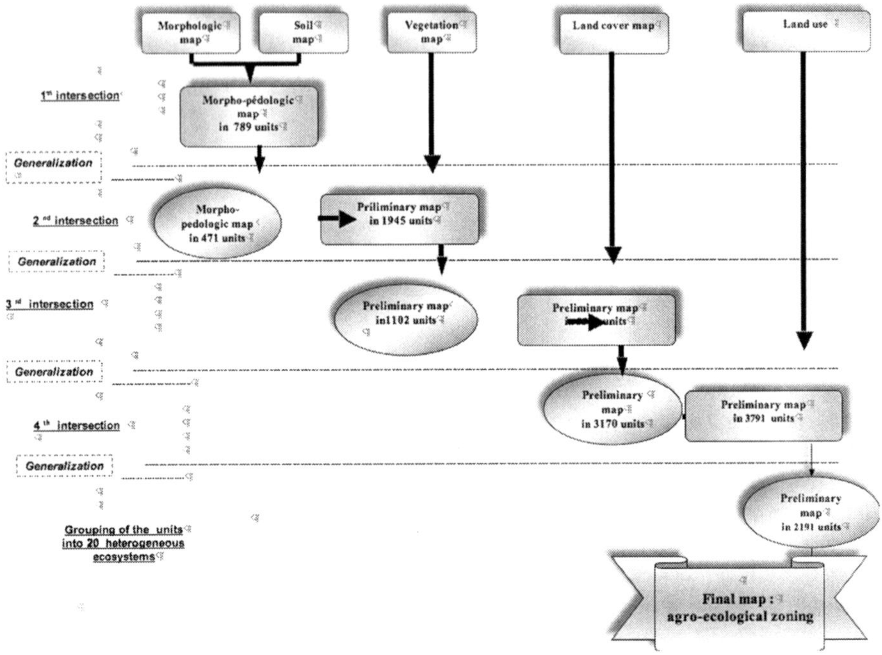

Fig. 1: Methodological approach

However, the intersection of the different cartographic layers is always a problem because it is very difficult to manipulate such amount of data. Taking into account these difficulties, in the beginning we acted by the intersecting the geographic layers two by two, using firstly the most important factors; secondly we did simplifications and generalizations of the result for better visualizations of the different maps.

3 Results

The final result of this work is an agro-ecological zoning map that shows 20 heterogeneous ecosystems. The legend of the map was adapted from Floret et al. (1978) (Fig. 2).

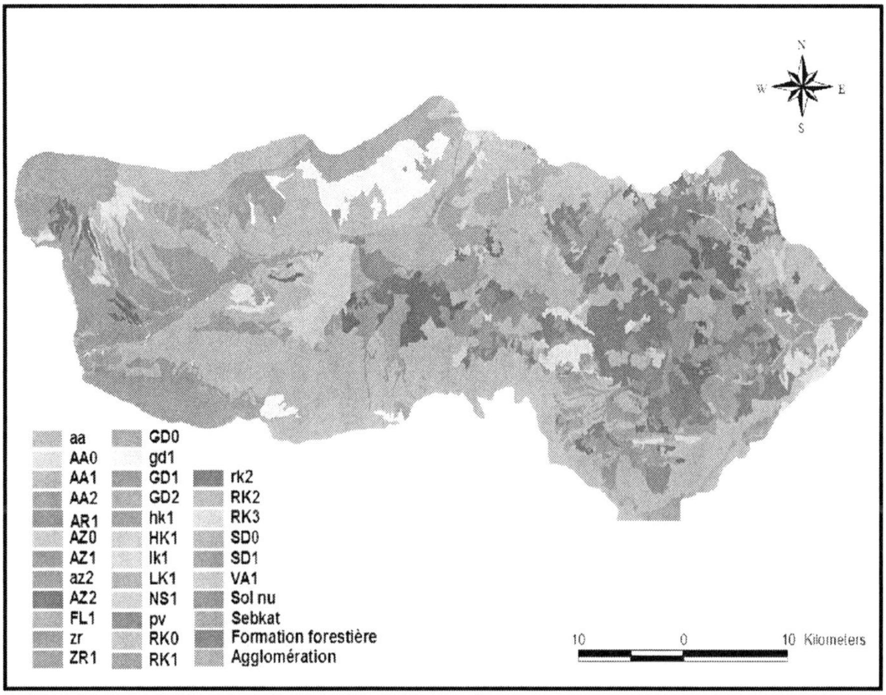

Fig. 2: Map of agro-ecological systems in Menzel Habib region

4 Perspectives

The development of this agro-ecological zoning offers new possibilities for the long term monitoring of environmental changes through:

Development of indicators of land vulnerability to desertification, so these indicators will help scientists to evaluate the state of natural resources and decision makers for a best management of environment within the framework of ROSELT/ OSS programme (Long Term Ecological Monitoring Observatories Network/ Sahara and Sahel Observatory).

Identification of early warning indicators of drought. This methodology will be realized in the framework of the SMAS project (Assessing drought in the Mediterranean) 2006–2010.

Bibliography

Ouled Belgacem A., Neffati M. 1996. *Etude de la dynamique de la végétation en milieu saharien* in Revue des Régions Arides-Médenine/Tunisie, 252–257.

Floret C., Le Floc'h E., Pontanier R., Romane F. 1978. *Modèle écologique en vue de la planification et de l'aménagement agro-pastoral des régions arides: Application à la région de zougrata.* PNUE/CEPE/ORSTOM/IRA/DRES. Document technique 2 (1978).

VERY HIGH RESOLUTION SATELLITES

LIVIO ROSSI
Eurimage S.p.A., via D'Onofrio 212, 00155 Rome, Italy
l.rossi@eurimage.com

Abstract

With very high resolution (VHR) satellite it is now possible to carry out, using collected data from space, many applications that in the recent past were exclusive to airborne and on site surveys. Due to the high geometrical resolution, the multi-spectrality, the high radiometric sensitivity, the revisit capabilities, the area imaged by a single frame or strip and to the accurate geometrical processing that can be targeted, VHR imagery is a key source of information for a wide range of applications related to natural, semi-natural, urban up to desert environment.

For example, a possible range of applications in this scenario can be: seasonal vegetation detection and monitoring, pasture monitoring and grazing paths, water points inventory, palm trees counting and monitoring, desertification expansion and sand mobilization mapping and monitoring.

The paper also gives satellite data technical features and in general the capabilities of Very High Resolution in territorial problems solving.

Keywords: Satellite, very high resolution, geometric accuracy, thematic detection, scale mapping, applications, desertification

1 Introduction

The use of Earth observation satellite data is growing in importance as new missions offer new types of data to new and expanding opportunities and markets.

With Very High Resolution (VHR) satellite, it is now possible to carry out, using collected data from space, many applications that in the recent past were exclusive to airborne and to site surveys.

Due to the high geometrical resolution, the multispectrality, the high radiometric sensitivity, the revisit capabilities, the wider area imaged by a single frame (respect to airborne) and to the accurate geometrical processing that can be targeted, VHR imagery is a key source of information for a wide range of applications related to natural, semi-natural and urban areas, including arid and desert environments.

In the past the use of satellite data (when limited to medium-resolutions of 10–30 m pixel size) was mostly confined to large, multi-national organizations that could support highly specialized applications and the highly specialized staff that

A. Marini and M. Talbi (eds.), *Desertification and Risk Analysis Using High and Medium Resolution Satellite Data*,

went with them. Medium resolution data still has an important part to play, especially for environmental monitoring and regional/national desertification analysis, but the arrival of very-high-resolution satellite data has meant an explosion of new applications and scale improving seeking new customers, as satellite data moves out of research laboratories and on to the desks of users with operational, real-world needs.

VHR satellite sensors can provide details, homogeneity, quality and timeliness, difficult to match with other remote sensing sources. The classic parameters of target analysis and interpretation can be in fact measured with VHR sensors:

- Tone and texture (rather than just pixels)
- Shape (e.g. for tree crowns, or small land parcels)
- Object distances, position and geometry
- Context analysis (e.g. different human activities)
- Multi-spectral analysis of objects rather than pixels or pixel groups

In addition, users can have (regional or local) information, which is continually updated, and can select the best acquisition dates for their applications and multi-temporal analysis from archives, which are growing continuously. Cartographic accuracy with 0.6–1 m of resolution goes up to 1:4.000/5.000, while thematic identification is possible at a scale of 1:1.000 or more.

2 The Very High Resolution sensors

The category currently includes the following satellites:

QuickBird, operated by DigitalGlobe™, distributed by Eurimage
Ikonos, operated and distributed by SpaceImaging
EROS, operated and distributed by ImageSat
Kompsat2, operated by Spot Image
Orbview (currently only from the archive)
Formosat2, distributed by SpotImage
Irs P5 Cartosat, distributed by Euromap

These satellites are quite dissimilar among them: EROS and Cartosat collect only panchromatic images, QuickBird, Ikonos and Kompsat2 use the same spectral bands, but QuickBird has a better geometrical resolution and a larger swath, Orbview (not currently operational) presented a small swath and, unfortunately, alternative acquisitions for pan and multispectral.

This huge quantity and quality of information available from space-borne sensors, allows applications ranging from international treaties to archaeology, from landmine clearance to tourist maps, fulfilling the whole requirements concerning natural, semi-natural and artificial areas. In desertification monitoring projects, VHR data appears very useful when within a synoptic frame of medium/low resolution imagery a sampling of key areas must be extracted, using the highest definition as ground truth, both for morphologic-morphometric detection and seasonal vegetation monitoring.

3 The VHR satellite features

For a correct comparison and selection of a VHR sensor we must consider: original digital data and the radiometric amplitude, resolution, multispectrality, swath, revisit time and, last but not least, offered products and services.

Very High Resolution data is therefore changing the field of large-scale applications and local, national and international Agencies/Organizations can plan ahead to include this tool to provide the best solutions to the user community (Google maps included).

Generally, digital sensors can improve both the reference scale of applications and thematic results through:

- Up to 0.6 m resolution in panchromatic mode at nadir
- Up to 2.4 m resolution in multi-spectral mode at nadir
- Simultaneous acquisition of pan and multi-spectral (ranging from 120–300 km^2)
- Radiometric range of 11 bit (2,048 levels of grey) instead of the usual 8 (256)
- Wider panchromatic band towards infrared (ranging from blue up to 1 μm).

The last two improvements, in particular, allow a better spectral and radiometric analysis, cleaner visibility of the images and therefore better discrimination, reducing mistakes, number of ground surveys and project costs.

In addition, it is possible to:

- Always have updated information
- Select the best acquisition date (e.g. for phenological seasons, water and humidity in environmental and desertification monitoring) for different areas and different purposes

Main requirements to be satisfied in order to use VHR data as a source of information for desertification monitoring can be identified as follows:

High geometrical resolution for large scale detection of morphological features and lithologic/geological aspects
Multispectral capabilities for rock and vegetation
Radiometric sensitivity for humidity detection and albedo measure and correction
Positioning accuracy for ground truth sampling
Revisit capabilities for local and adequate information frequency
Image size suitable to tailored study/projects

In addition, the best VHR highlights can be summarized as follows:

Geometrical resolution – the native resolution when collecting at nadir (e.g. Quickbird has 0.61 m that becomes 0.66-m when collecting with an acquisition angle of 15°). When the collection angles are in the range 0°–15°, the geometrical resolution can be easily compared with the one resulting from an average scale aerial flight.

Multispectrality – the best sensors collect the frame on a higher resolution panchromatic band extended also in the near infrared, and on four multispectral bands (three in the visible, one in the near infrared). By using data fusion techniques, the multispectral bands can therefore be merged with the panchromatic, in order to obtain the best high-resolution pansharpened, both in natural and in false color.

Radiometric sensitivity – When the dynamic range has 11-bit (2,048 levels of grey), imagery can be better stretched specifically, in order to improve the visual information. This feature is very useful when dealing with mountain-shadowed areas or with albedo high reflectivity (desert/glacier environments).

Positioning accuracy – Data must be geometrically processed with different techniques for easily inserting into a GIS through:

Ortho-rectification with a rigorous sensor model
Ortho-rectification with the RPC (Rational Polynomial Coefficients)

The first approach is possible when the basic product packaging includes not only the image file, but also a list of ancillary data files, such as those containing ephemeredes and sensor parameters. These data, managed by proper software that models the satellite sensor, enable the generation of accurate ortho-images. Software packages that can handle the rigorous sensor models are available on the market.

Revisit capabilities – Satellites have a reviewing rate that depends on the off nadir angle. In any case, applications that require multi temporal observation with seasonal frequency (e.g. desertification monitoring) can be easily carried out.

Image size – When collecting with an off nadir angle in the range 0°–15°, the image covers approximately a swath with a difference of more or less 10–15% (e.g. Quickbird, in the swath range from 16.5 to 18.0 km brings the extension of each frame from roughly 270 up to 300 km^2. Given this frame extension, is possible to plan the useful numbers for each project, including eventual overlapping.

4 Main sensor characteristics

IKONOS

Launched the 1999 September 24th, was the first commercial that allowed the imagery acquisition of 1 m resolution.

Orbit:	Polar sun-synchronous
Altitude:	680 km
Frequency (theoretical, minimum):	14 days
Frequency (potential max):	2.5 days
Swath:	11 km
Bands number:	4 + PAN
Spectral resolution:	0.45–0.90 μm
Spatial resolution:	Panchromatic 1 m Multispectral 4 m

QUICKBIRD

Launched in 2001 October is currently the satellite with the best spatial resolution for civilian uses.

Orbit:	Polar sun-synchronous
Altitude:	450 km
Frequency (theoretical, minimum):	14 days
Frequency (potential max):	3.5 days
Swath:	16.5 km

Bands number: 4 + PAN
Spectral resolution: 0.45–1 μm
Spatial resolution: Panchromatic 0.6 m
Multispectral 2.4 m

EROS 1&2
Launched on 2000 and 2006 by Israel.

Orbit: Polar sun-synchronous
Altitude: 500 km
Frequency (theoretical, minimum): 14 days
Frequency (potential max): 2.5 days
Swath: 12.5 km
Bands number: 1
Spectral resolution: Only visible
Spatial resolution: Panchromatic 1.8 and 0.7 m

ORBVIEW 3
No more operative, to be checked for the archive.

Orbit: Polar sun-synchronous
Altitude: 470 km
Frequency (theoretical, minimum): 14 days
Frequency (potential max): 3 days
Swath: 8 km
Bands number: 4 + PAN
Spectral resolution: 0.45–0.90 μm
Spatial resolution: Panchromatic 1 m (pan or multispectral, never together). Multispectral 4 m

FORMOSAT
Launched on May 2004 by Taiwan.

Orbit: Polar sun and geo-synchronous, only over the same word strip areas
Altitude: 891 km
Frequency (theoretical, minimum): 4 days
Frequency (potential max): 1 day (only over selected same areas)
Swath: 24 km
Bands number: 3 + PAN
Spectral resolution: 0.45–0.90 μm
Spatial resolution: Panchromatic 2 m
Multispectral 8 m

KOMPSAT
Launched on 2007 by South Korea.

Orbit: Polar sun-synchronous
Altitude: 685 km
Frequency (theoretical, minimum): 12 days
Frequency (potential max): 2 days
Swath: 15 km
Bands number: 4 + PAN

Spectral resolution:	0.45–0.90 μm
Spatial resolution:	Panchromatic 1 m Multispectral 4 m

5 Applications and services

The following list gives examples of the most important application opportunities for civilian uses, using Very High Resolution Satellites (Figs. 1–8).

Agriculture

Production status and trend (inventory and yield estimation)
Production process management (field variability and crop vigor)
Subsidy area control (field measurement and single crop identification), even to the point of almost avoiding ground surveys
Agricultural cadastral parcel updating (local taxes and subsidies)
Updating of permanent crops registers (olives, vineyards and citrus)
Vineyard trend analysis and harvest management
Crop damage assessment and mapping at parcel level, coming from hail, rain, pests, drought, wind storms
Irrigation efficiency

Forestry

Structure/typology maps
Illegal logging (forest losses due to illegal harvesting)
Local health analysis (up to single crown)
Fragmentation index and mapping
Contribution to fuel maps at large scale
Health evaluation of surviving trees (post crisis/fire)

Cartography/cadastre

DEMs and city models (3D)
Updating topographic maps (object positioning up to 1:4.000; thematic polygons up 1:1.000)
Identifying cadastral parcels to be updated by traditional means (if the scale is larger than 1: 4.000)
Illegal building detection (not accessible or hidden zones)
Infrastructure detection and monitoring (down to telephone/lighting poles)

Environment/water

Fire impact evaluation for sensitive habitats and species
Desertification analysis and monitoring
Local wildlife habitat maps
Freshwater and pollution in coastal ecosystems (11 bit permits penetration of shallow water)
Coastline condition and changes (beaches, submerged vegetation, urbanization, oil on the shore, etc.)
Coastal sediment transportation analysis

Aquaculture and shallow water monitoring
Inland water/wetland monitoring
Waste deposit and quarry detection and monitoring
Identification and quantification of air (sometimes water) pollution sources

Land/urban management

Land use mapping at regional/local scale (thematic identification up to 1:1.000 scale)
Thematic and geometric support to Corine Land Cover up to 4° level
Soil composition and structural morphology (limited areas at large scale)
City parks management and monitoring (at single tree level)
Contribution to design (and check) of telecommunications networks (antenna positioning, wireless and cellular networks)
Contribution to public services inventories

International treaties, public safety and health

Clear picture of situation for areas of limited media access (or denied areas)
Compliance verification for international agreements (site inspection, etc.)
Contribution to de-mines projects
Information for humanitarian emergencies, tracking refugee movements
Resource availability information for humanitarian response teams
Management of industrial disasters

Risk management

Flood boundary definition at large scale, for flood damage assessment
Assessment of single landslides (without stereoscopic view) and natural damage (earthquakes, storms, floods, etc.); monitoring down to single building level
Disasters management preparedness and mitigation efforts

Transportation and services

Transportation plan preparation (corridor studies: highways, railways, pipelines, etc.), road networks assessment and monitoring
Electricity, telecom, water, oil and gas transmission and distribution corridors assessment and monitoring
Road network extraction and evaluation (status, changes, markers, etc.)
Digital database (e.g. TeleAtlas, Navtech) overlap for updating and completing information

Archaeology

Buried structures, excavation updating
New archaeological plan, tourist parks

Advertising and media

Education and web
Tourist maps and facilities; posters
On-line maps background (e.g. Google maps, etc.)

6 Conclusions

The optical VHR satellites appear as the tools that best fit the above application needs and requirements and, due to their geometrical resolution, are the best choice (also airborne data, but for regional acquisitions) when a high detail is required. In order to allow the extraction of layers for a GIS, the ortho-correction of the imagery is obviously mandatory. By using commercial software, it is possible to reach accuracies even in the order of less of 1-m RMSE. After the ortho-rectification phase, a quick image enhancement session, carried out with entry-level image processing software, enables the use of the satellite data within a GIS environment for the needed information extraction, both geometric and thematic.

In general, the best situation for a complete detection of the above described targets, is to start from pansharpened images (natural or false color data fusion at the pan resolution) collected from March to November with an acquisition angle below 20°, even if in desert environment the acquisition period, due to the more acceptable sun elevation can go up to around the year. The reachable thematic scale is depending on the requested specifications and geometric accuracy of each sensor.

With the new imagery approach (e.g. Google Earth maps) the large-scale knowledge appears now mandatory, both for technicians and consumers. With the VHR images "era" the planet investigation has become possible and accessible, giving a complete and updated geographic prospect to everyone.

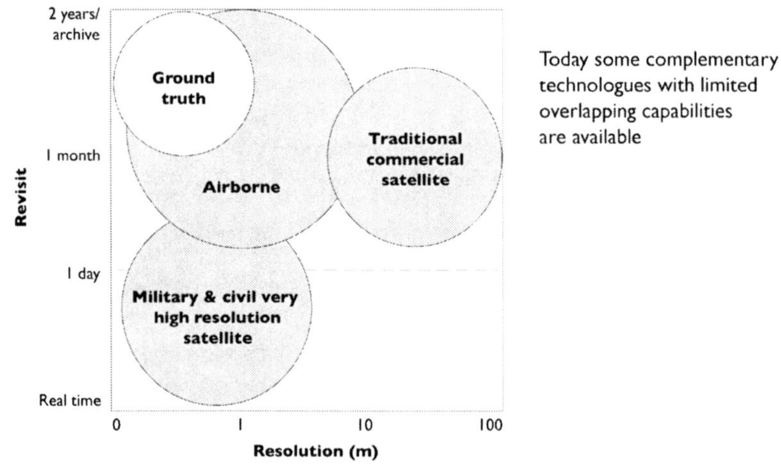

Fig. 1: Comparison between the average time of revisit and the resolution of the main remote sensing groups

Fig. 2: Comparison, at the same high enlargement, between traditional analogue remote sensing data (left) and digital Quickbird imagery (right)

Operational VHR satellite comparison

	DIGITALGLOBE	SPACE IMAGING	ImageSat International
Orbit	450 Km	600 Km	480 Km
Resolution at nadir	0.61 m	1 m	1.8 m
Swath frame	16.5 Km	11 Km	12.5 Km
Memory on board	128 Gbit	64 Gbit	3-4 scenes
Bands	PAN + 4 MS	PAN + 4 MS	PAN
Acquisition frequency (off nadir)	5 days (0.61 m res.)	4 days (1 m res.)	3 days (1.8 m res.)
Minimum order for new acquisition	64 Km^2	100 Km^2	On request

Fig. 3: Example of characteristic comparison between the first available sensors

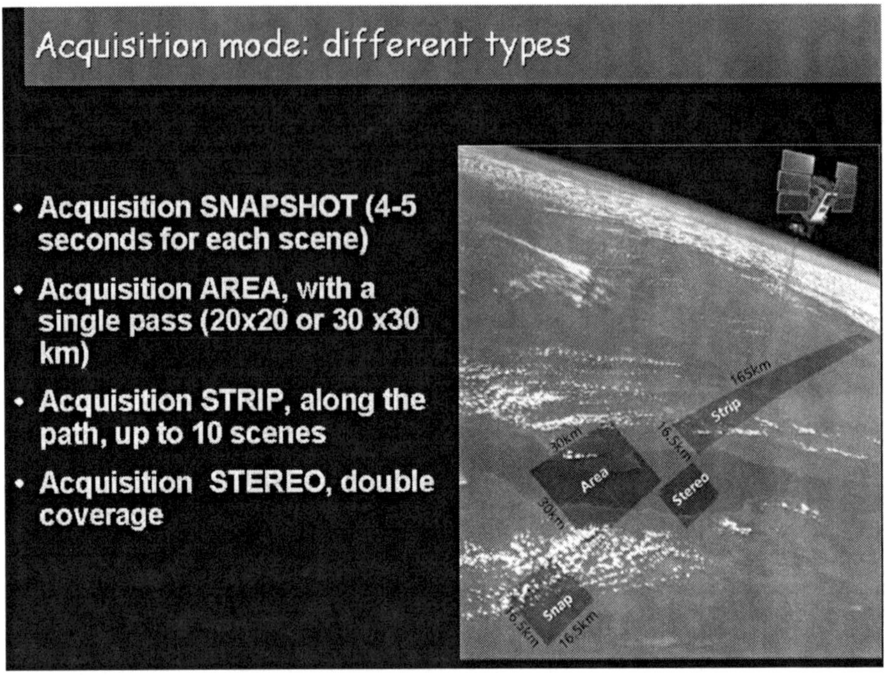

Fig. 4: Example of VHR satellite acquisition mode

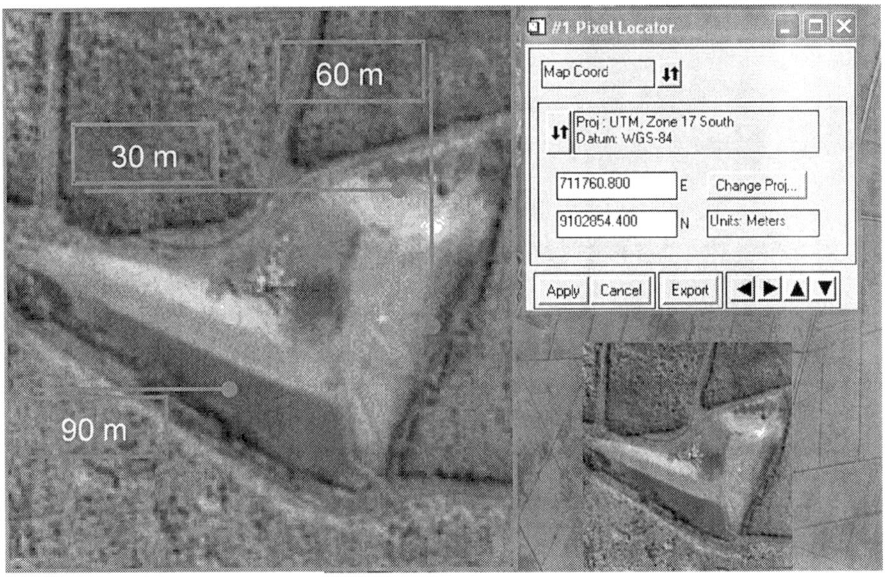

Fig. 5: Application in archaeology. All over the world, detection and measurements of buried finds show the useful capability of the VHR sensors

Fig. 6: Monitoring of oil extraction structures in arid areas

Fig. 7: Desertification analysis and monitoring

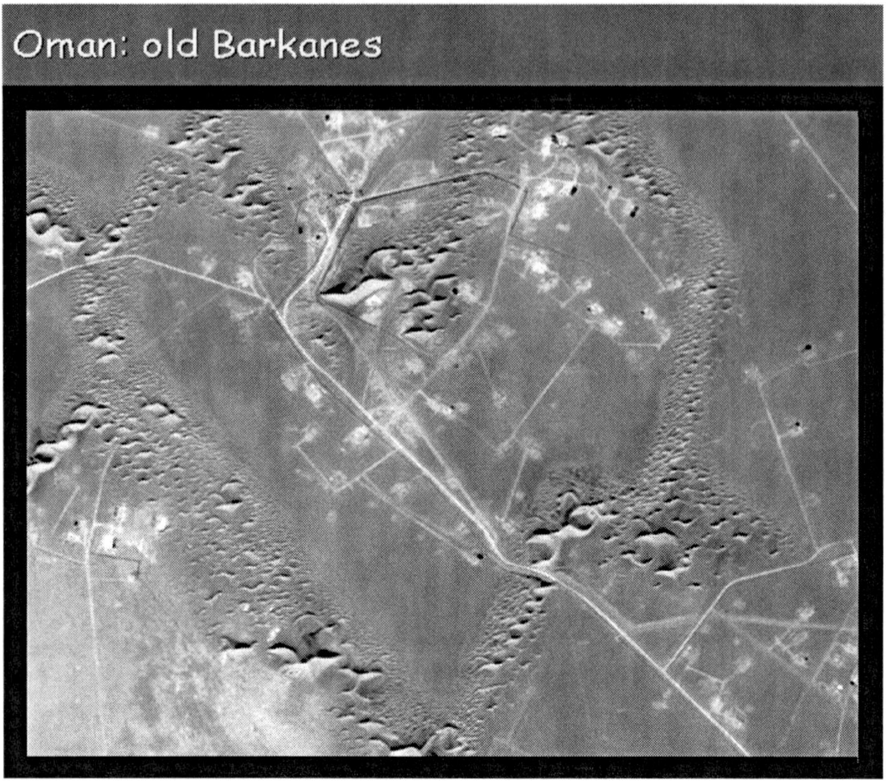

Fig. 8: Morphological and morphometric analysis by VHR data

VULNERABILITY OF THE LOWER SAHARA AND THE HUMAN IMPACT ON THE INCREASE OF DESERTIFICATION (*THE CASE OF OUED RIGH AND OUED SOUF REGION*)

LAKHDAR MERABET
University Kasdi Merbah Ouargla (ALGERIE)
merabetlakhdar@yahoo.com

Abstract

The increased exploration of the underground water and soil in the Northeast of the Algerian Sahara for the sake of drinking and irrigation has resulted in catastrophic hydro-geological and environmental problems, especially the rise of the level of the superficial layer that affects agriculture and environment. This is mainly a consequence of the mismanagement of the extracted water resources, that has in turn caused bio-climatic and geomorphologic problems, especially in the Oued Souf and Oued Righ region. This phenomenon has escalated, especially after the increase in the land cultivation projects, which were often practiced without prior hydro-cultural analyses, as well as a consideration for irrigation water drainage. This has in turn created other problems, such as a significant increase of the amount of salt in soil and which has resulted in the spoiled and desertification of large fertile plots of land. Besides, the aggressive livestock herding causes poor biological renovation of the wild plantation.

In addition, there are some social communities that consume high amounts of water, which exceed 500 L per day per capita, while the sewage system is underdeveloped and cannot cope with such high levels of drained water. As a result, water is drained in some traditional wells (septic pits). The poor management of water resources resulted in disastrous outcomes, which consisted mainly in the rise of the free level of the phreatic water and saltiness, which affect agriculture and infrastructure. Thus, this paper aims at shedding more lights on the desertification and degradation of the environment, as well as compiling additional scientific proofs for the research dealing primarily with the results of the human interference in the environment.

In order to lessen such catastrophic results, a reasonable management of underground water resources cannot be neglected, especially in regions supplied with such a strategic resource, for a sustainable development. Otherwise, the escalating mismanagement of our water resources, with no planning and policy, would lead to disastrous outcomes. A continued disrespect of this precious resource means an unwise consideration of the fate of the future generations. Therefore, directory plans for the arrangement and management of the soil and water resources, in terms of hydro-agricultural capacities, should be put in place. In addition, regular measuring

A. Marini and M. Talbi (eds.), *Desertification and Risk Analysis Using High and Medium Resolution Satellite Data*,

of the piezometric levels of different underground napes exploited should be recorded. The technical outline of research in the hydro-agricultural and urban arrangement fields should be identified before hand. This would improve the management of soil and water resources, as well as stimulating search for irrigation systems suitable for the climate conditions of the region.

Keywords: Lower Sahara, environment, salinity, rising water, irrigation, desertification, vulnerability

1 Introduction

Dawn in history, so many well-known civilizations existed respectively on the two edges (sides) of the lower Sahara Rivers: Oued Souf and Oued Righ. In fact, they do not currently represent rivers by their geomorphological conception; but they are rather names of geographic regions. Nevertheless, one could not deny the fact that they are inherited rivers; in that the region has seen very rainy periods, especially during Pleistocene 4th geological age. Oued Righ for example, is generally situated in a lower region; its peak is 103 m high, it is located in the south (Goug). The river flows towards the north and falls in Oued Khrouf which is 130 km far from. The latter flows and falls in the lake of Marouan and from there in the lake of Malghrigh.

It is important to mention that the basis of the ancient civilizations used to be the use of traditional irrigation means in hard and draughty environmental circumstances, despite the fragile environmental system. Mankind was able to keep balance in nature with his interference, by using only the superficial layer, which is found at a maximum depth of 60 m. Thus the water extracted from the layer used in agriculture and other human uses is turned back to the same superficial layer diminishing the amount of consumed or vaporized water. The last century has seen an increase in digging and drilling for water wells in different deep layers, in addition to the over-exploitation of the under water in unprecedented ways. For instance, in 1956 the number did not exceed 300 wells. However, it has jumped to 1,033 in 1922 for the irrigation of 300,000 and 600,000 palm trees in the region of Oued Reigh. These new drilling operations and that huge number contributed to the drought of the superficial layer. This phenomenon simply expresses the degradation of the region from Sidi Rached until Ghuemra Oases. The region's superficial layer has been greatly degraded, which has affected the neighboring Oases like Tamarna region.

The different types of human activities do not affect greatly the rich environment where the rich land and the biological variety help in the innovation process, whereas human interference could lead to catastrophes in droughty regions such as the region under study here. The final consequence of human interference is the spending of large sums of money in what is called the cultivation and urbanization of the region and their subsequent negative results.

In this lecture, it is necessary to have a geographical overview of the region before listing the most important example that explains this problem.

2 Methodology

In this study we depend on the inductive method; the project is therefore divided into three steps: the first step is the Documentary Phase in which we determine physically the study field and the notions that are linked to the project through studies, reports, maps and aerial photos that have been done on the region. In the second step all of the statistics linked to the project from different technical and academic services are gathered; in the third step all the information gathered above, in addition to the field work, where measures have been taken which help in counting the speed of desertification and observing the symptoms of the physical field degradation, are analyzed.

3 Location of the region

The region is astronomically and geographically situated between Eastern longitudes 4°–8° and Northern latitudes 32°–34° (see Map No. 01).

The region is bounded by the Aures Series Mountains from the North, the Great Erg from the East, the hill of Tadmate from the South, and the heights (platform) of M'zab from the West. As the region is divided geographically into Oued souf and Righ, it is also divided administratively between Ouargla and Oued Souf (see Map No. 02).

The geographical study location of the region has great importance because of its direct impact on the climate of the region in addition to other factors such as topography and hydrology, etc.

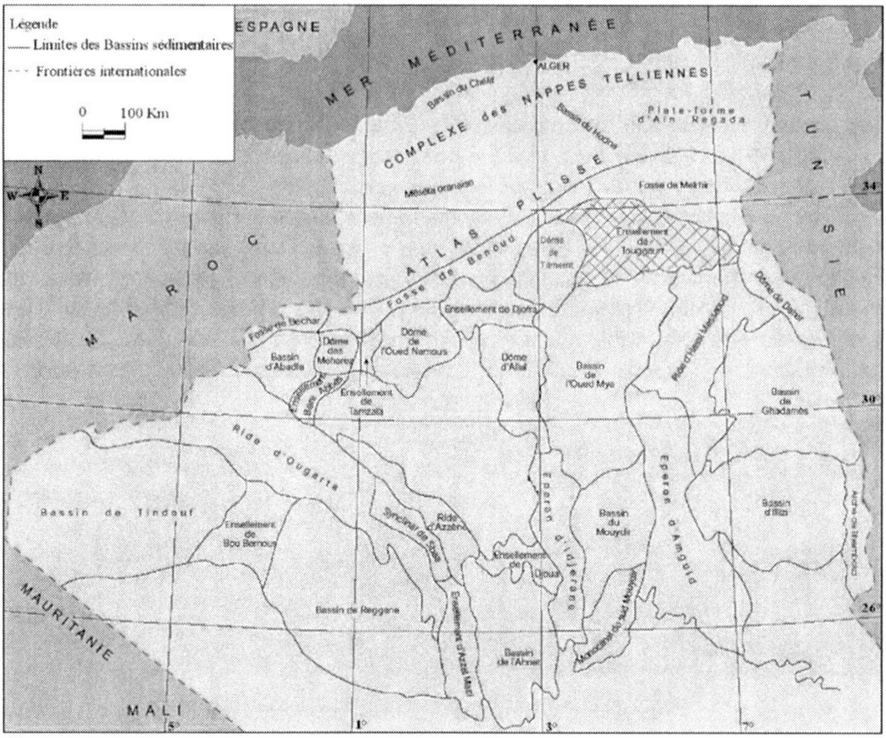

Map No. 01: Geographic location

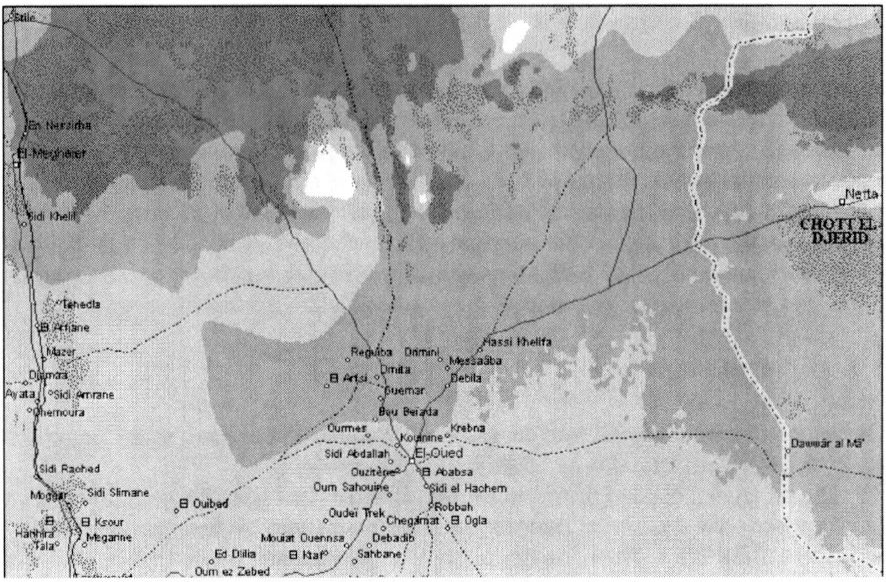

Map No. 02: Details of the region

The paper does not allow here to elaborate on the above points; therefore, focus will be here on the degradation and desertification of the region. In fact, the causes of desertification are classified into natural and human.

4 The natural causes

The natural causes can be summarized in the different drought figures which characterized the region in the Holocene period, especially before 5,000–6,000 years. So let's have a look on the climate of the region (Figs.1 and 2).

The interaction of the climatic elements is clearly seen in the lower Sahara; each element is linked to the other either physically or by affecting the physical medium. The climate characteristics, especially the high temperature, humidity deficiency, and high evaporation of precipitation (rainfall) affect the plants' need for water. This is of course followed by the extent to which the plants are affected by the additional salt concentration in the soil water due to increasing evaporation rates. Consequently,

Figs. 1 and 2: Inherited aspects of Sahara (region of Touggourt)

and over time, the plants' absorption of salt increases and its productivity decreases and hence the amount of money spent exceeds production. We therefore come to the conclusion that hydraulic systems and their relation with the climate is important.

In order to obtain an objective idea about the climate of the region we have chosen the climatic data provided for long periods, beginning with Seltzer data from 1913 to 1938 and Gean Dupief data from 1926 to 1956, in addition to the meteorological station data in Gamar and Sidi Mehdi from 1970 to 2004.

Generally the region is characterized by yearly average temperature of 22° whereas the precipitation (rainfall) average is less than 100 m.

The evaporation is directly affected by the high temperature and relative humidity (Table 1).

Table 1: Monthly evaporation average in mm during the period from 1980 until 2002 in both sidi-mahdi and gmar stations (Gmar and Sidi-Mahdi Meteorological Stations 2003)

Sep	Oct	Nov	Dec	Jan	Feb	Mar	Apr	May	Jun	Jul	Aug	Total
218.05	148.74	99.68	77.57	82.87	95.8	128.28	191.33	247.75	281.09	305.79	281.23	2158.8

Evaporation average in Souf

Sep	Oct	Nov	Dec	Jan	Feb	Mar	Apr	May	Jun	Jul	Aug	Total
297.43	209.61	134.87	116.4	112.4	146.22	227.13	282.13	357.3	417.1	440.95	407.82	3149.16

Evaporation average in Righ

Examining the data in the table above gives the initial perception that there has been a huge amount of water evaporation in the air, as measured at the meteorological station. However, in reality this amount does not exist in the Sahara environment so that it could vaporize simply because the perception is deficient, thus I count the real ETR evapo-transpiration by Turc's equation (Table 2).

Table 2: ETR values during the period from 1970 until 2002 (Counted by the researcher according to TURC's equation)

Month	Sep	Oct	Nov	Dec	Jan	Feb	Mar	Apr	May	Jun	Jul	Aug	Total
ETR(Souf	5.06	7.38	10.22	7.45	12.75	8.54	9.70	6.75	5.37	1.79	5.27	1.05	81.36
ETR(Righ	4.41	5.07	8.74	6.24	10.41	9.91	9.21	6.14	4.78	1.88	0.61	0.86	65.26

5 Winds

Winds play an important geomorphological role particularly in the droughty, semi-droughty and coastal region. In fact, the principle morphological features in the region that are constituted by winds reflect clearly its characteristics in direction, speed and frequency. Therefore, the study of the forms resulting from winds and air operations becomes more important by repeated measuring of the winds speed, frequency and directions. Generally, the frequency speed in the region is slow, but it is more effective in April and July when the winds become stormy, especially in the East and Southeast (Table 3).

Table 3: The average speed of winds in m/s during the period from 1980 until 2002 in both Sidi-Mahdi and Gmar stations (Gmar and Sidi-Mahdi Meteorological Stations 2003)

Month	Sep	Oct	Nov	Dec	Jan	Feb	Mar	Apr	Mai	Jun	Jul	Aug
Sidi Mehdi S.	2.71	2.65	2.73	3.83	2.56	2.73	2.26	3.93	4.61	3.66	3.26	2.93
R. Gmar	3.04	2.53	2.50	2.40	2.82	3.14	3.65	4.38	4.66	4.77	3.83	3.38

The above table confirms and demonstrates what has already been explained. It is clear that wind speed in El-Oued is faster than in Oued Reigh. It is between 4.38 and 3.83 m/s in April and July, the season of the strong winds. In Oued-Reigh, the average of the winds in April is between 3.93 and 3.26 m/s in July.

In general, the winds frequency in Oued Sauf blow from East and North–East, which are the dominant winds, as well as the seasonable southern winds, called locally the Shelly winds, which blow at a frequency of between 4.54 and 2 m/s. (Table 4):

Table 4: The annual average of winds days according to their directions, during the period from 1980 until 1999 in Gmar station (Gmar Meteorological Station 2002)

Direction	North	NE	E	SE	S	SW	W	NW
Days number	40	45	75	35	13	40	23	43

The study of the winds in the region is important because of the vulnerability of the region, especially in the movable sand regions. The winds displace the sand from one place to another, which leads to the covering and desertification of the agricultural fields. It also covers the buildings and roads after every strong wind causing the isolation of the region for many days. These catastrophes inspire people to install obstacles and barriers in an effort to break the movement of the sand (Figs. 3 and 4).

Figs. 3 and 4: Barriers for breaking sand movement (region of EL Rebbah)

6 The human causes

The human causes of degradation and desertification of the physical medium can be summarized in human interferences either in agriculture, industry or services, which do not respect environmental balances, especially in the vulnerable regions. Human impact is clearly seen in the services factor by establishing different infrastructures (roads, railways, gas pipelines, etc.), in addition to tourism activities. These activities transform the stable sandy dunes into movable sands in case of blowing winds. In this paper, I will focus on human interferences through pasturage and water extraction.

7 The hydro-geological features of the region

Hydro-geologically the region is well known thanks to the studies carried out. In addition to that, many drilling operations were performed either by Hydro-Directory, Hydro-Drilling or Sonatrach (national oil company)

Generally, the lower Sahara region is situated in a big sedimentary basin, which constitutes a vast hydrological basin too; its surface covers 780,000 km^2, its thickness is between 4,000 and 6,000 m according to the UNESCO study. Therefore, it assimilates one of the biggest underground water resources in the world. The basin stretches to 700,000 km^2 in Algeria and 800,000 km^2 in Tunisia. The basin's layers appear at the Atlas Saharan series in the North. These boundaries are characterized by so many faults, whereas from the West, we find El Hadjri line and Ouad Essaoura, Tadmait which stretches towards the East-West from the South, the region of Kabiss and cretacy elevations in the mountain of El Dahir from the North East of the basin, and the political boundaries from the East (Table 5).

Table 5: Hydro geological section towards NW/NE (UNESCO 1972)

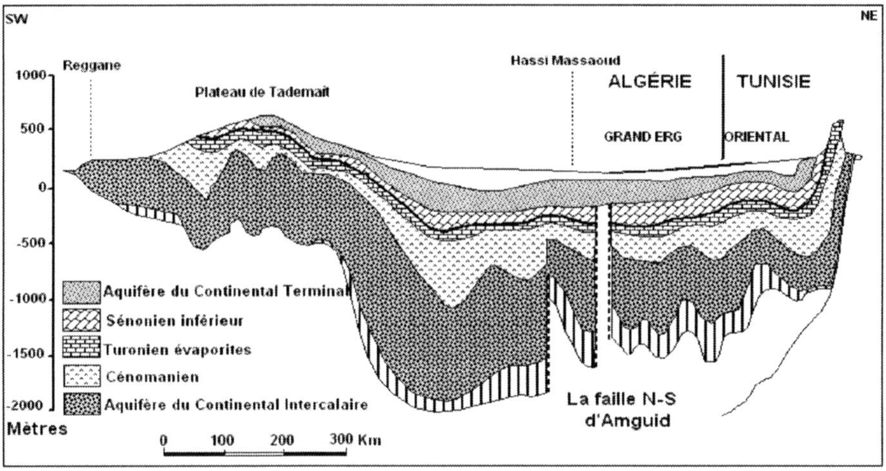

The problem lies in the fact that the extracted water exceeds the average of the water layer nutrition. Generally, in this vast basin we can distinguish two aquifers, which are characterized by their independent hydrodynamic conditions.

8 The aquifer continental intercalary system

It is the biggest water layer in the lower Sahara. Generally, the basin is situated in continental constitutions of the inferior layer. It is formed of stone, clay, and grain, its depth changes from one region to another. For instance, in Ouargla, it is 1,058 m deep, in Djamma, 1,515 m, and in Blidet Amor 1,650 m. It seems that the aquifer depth decreases whenever we come close to the resonators that are characterized by hot water.

In general water in continental constitutions is hot. It is over 60°C alpine and Barmian layer. It is also lowly salt reaching 2–4 g/L, therefore, it is different from the terminal complexion (aquifer). This aquifer covers a vast surface of about 850,000 km². It is characterized by three partial layers: Synonion/charbonic Iosan and Miupiusan. It is relatively highly salted. It is about 70–280 m deep. The flow is approximately 30 l/s (Figs. 5 and 6).

Figs. 5 and 6: Water pressure in continental intercalary and terminal complexion

9 Water quality and its effects on desertification

The study of water quality becomes very important according to its direct impact in changing the irrigated lands into desert ones, due to the accumulation of salt melted in this water, particularly in the hard climatic conditions, high temperature and evaporation. In addition to the geomorphologic cause, hydrological, Topographical and human causes have to be taken into consideration. It has been confirmed that the region's sole contains very concentrated salt. This is clearly seen in the existence of superficial salty crust, which differs in its components and appearance according to the dominant type of salt in the soil, amount of the organic substance and the soil humidity degree. The humid dark crust reveals that chloride, calcium and magnesium chloride are tense in the soil; Whereas, on the surface a soled crust is formed with the existence of gypsum, and the concentrated salt of $NaSO_4$ with plaster and $CaCO_3$ give a soft powder crust on the surface of the soil (Figs. 7 and 8).

Figs. 7 and 8: Salts sedimentation near Sebkhas due to evaporation

The most sediment salts are: $CaCO_3$ at 88.75%, $CaSO_4$ at 3.25% and $MgSO_4$ at 6.05%.

One of the causes of salt accumulation in the soil is the water extracted from the aquifer continental intercalary system and complex terminal. In fact, this extracted water does not affect negatively only on the soil, but also the hydraulic infrastructure, by diminishing its prospective age. Some of the melted salts in the extracted water sediment in these infrastructures (Figs. 9–12 below).

Figs. 9, 10, 11, 12: Salts sedimentation causes the degradation of hydraulic constructions

In fact, these salts result from the rocky composition of the aquifers. This can be seen by examining the stratigraphic unit.

The extraction of large amounts of water by mankind is the only cause of desertification, but one has to take into account also aggressive pasturage in the region.

10 Pasturage as a cause for desertification

Indeed, pasturage is one old function in the region, in that the region used to enjoy a rich biological variety much better than today: the existence of a zoo in the region, full of different kinds of local animals, from deer to rabbits, supports this opinion. However, many of these animals are threatened by extinction because of the man's interference either by hunting or pasturing, in that pasturing by the region's Bedouins camels and livestock strained the pastures for which other wild animals had to compete for surviving (Figs. 13–16).

Figs. 13, 14, 15, 16: Pasturing: humans contribute also in desertification

11 The animal resources in the region

Grazing is considered one of the most important jobs in the region in addition to agriculture. It can be divided into two kinds:

(1) It depends on stall feed is brought to their stables.

(2) It depends on pasturage in open fields. It includes cattle, goats, camels and cows.

(A) Goats are the most numerous in the region of Oued Reigh. There are about 98,815 heads, they are over 52.03% of the total livestock. The greatest percentage of them is in Jaama with 22.60%. The remaining is divided between Temacine, Touggort and Maggar. In Oued Souf the number of goats is higher; it is over 183,620 heads.

(B) Sheeps are second in number after goats with 80,328 heads and a percentage of 64.54%; they concentrate in the north of the region because of the green fields for pasturage. The largest number of these sheep is in Oued Souf; it has over than 90,256 heads, whereas Touggourt has a lesser number with 7,255 heads. Generally, Oued Souf is more pastoral than Oued Reigh.

(C) Camels: Ever since, camels have been linked to the region although their number is smaller than goats and sheep. There are around 6,213 heads with a percentage of 1.38% and 2,307 heads in Oued Souf. Oued Souf has a lesser number because of its worse pasturages. Generally, pasturages are poor in both regions because of the extensive exploitation either by man or animals.

(D) Cows: Cows are new in the region and so are their pasturages. As a result, their number is still low. Moreover, the cows' hay is too expensive.

There is still another problem in the region, which is the existence of huge numbers of livestock that invade the region's pasturages during fertile seasons.

12 Conclusion

The aim of this paper has been to summarize the natural and human causes of desertification in the region of the lower Sahara. In order to mitigate the disastrous consequences of the desertification phenomenon the following recommendations should be taken: first, launching an awareness-raising campaign in order to explain to the public administration, as well as the region's inhabitants, the vulnerability of their environment, so that they can modify their style of interference. Secondly, to rationalize water consumption for irrigation by observing hydro-agricultural standards, especially in soil salt cleaning and drainage, irrigation dose, irrigation techniques, as well as irrigation frequency. Thirdly, to build depuration stations for sewage water. Fourthly, periodic measuring of saltiness levels by physio-chemical analysis of soil. Fifthly, to regulate pasturage by protecting the natural biological renewal of the vegetation. Finally, to use remote sensing technology for monitoring the most vulnerable regions to desertification and to update the maps.

Bibliography

Agence nationale pour l'aménagement du territoire. 1995. *Plan d'aménagement de la Wilaya de Ourgla.* Rapport d'orientation 1995.

Agence de bassin hydrogrphique Sahara. *Apercu sur la pollution D'eau Existante dans la region d'Eloued.*

Nedjah Ahmed. 1971. *Le souf des oasis edition la maison de livres alger.*

BRL Ingénierie and Bneder. 1999. *Etude du plan directeur général de développement des régions sahariennes ressources en eau: connaissances d'ensemble.*

BRL Ingénierie and Bneder décembre. 1998. *Etude du plan directeur général de développement des régions Sahariennes – dossier de ressources en sols.*

BRL Ingénierie et Bneder. décembre octobre 1999. *Etude du plan directeur général de développement des régions Sahariennes programme, calendrier et conditions de mise en ceuvre « Commune de Touggort » 1978 Plan d'urbanisme – Directeur Phase I et II : Situation actuelle et perspectives – Juin 1978.*

Bureau National d'Etudes pour le Developpement Rural et DSA Ouargla. November 1992. *Etude d'un schema directeur de mise en valeur agricole de la wilaya de ouargla Mission étude Préliminire, Annexe n°1 Etude climatique.*

A. Cornet. and Rognon. Phd, *Estimation de la valeur des débits circulant dans la nappe du Continental Intercalaire au Sahara sud –algérien.*

Cornet. 1964. *Introduction à l'hydrogéologie du Sahara.*

A. Salem consultant – mars 2001 *La demande en eau dans le sahara septentrional algerien.*

Pierre Bellair. 1940. *Les sables de la dorsale saharinne et du bassin de L'oued R'hir* Imp. la typo – litho et jules carbonel réunies 2, rue de normandie, 2 1940 Alger.

G. Busson. 1970. *Le Mésozoïque saharien* (essai de synthèse des données de sondages algéro-tunisiens. Tome **2**).

Jean Chalie. 1972. *Le quaternaire l'histoire humaihe dans son environnemnt* éditeurs 8, place de l'odén – paris.

Marc Cote. 1999. *Macrocéphalie et micro – urbanisation, Cas du Sahara algérien: L'architecture et la ville face aux défis du 21 ème siècle* Séminaire Biskra 20 et 21 november 1999.

Direction de hydraulique de Ouargla. 2003. *Rapport sur la situation de drainage et d'evacuation des eaux D'assinssement de la cuvette de Ouargla* Iltalconsult – Rome, septembre 1970 amenagement hydro-agricole de la plaine D'abadla.

Jean Dubief. 1959. *Le climat de Sahara*, Tome: **1 and 2** Institut de météorologie et de physique du globe de l'Algérie. Alger 1959.

Jean Dubief 1963. *Hydrologie de Sahara.*

Daniel Dubost. *Ecologie d'aménagement et de développement agriculture des oasis Algériens.*

Philippe Duchaufour. 2001. *Introduction à la science du sol, végétation, environnement* 6e édition de l'abrégé de pédologie 2001.

Jean fabre 1976. *Introduction a la geologie du sahara algerien et des regions voisines.*

A. Farhi, Mazouz, S. Alkama, Naceur, Az. Saouli. 2000. *La wilaya d'eloud ou le probleme des disparites communales*, in Journal Algérien des Régions Arides.

Masson and Cie. 1967. *Précis de biogéographie* imprimé en France par brodard-taupin imprimeur-relieur coulommiers-paris 64175.

Jean – Michel Freulon. 1964. *Etude géologique des séries primairs du sahara central (tassili .n'Ajjer et fezzan)*, publications du centre de rechereche sur les zones arides centre national de la recherehe seientifique série: géologie.

Abdelmalek Nemouchi. 2001. *Geograpie hydrologique du bassin versant endoreique du chott el hodna, these présentée pour lobtention du dilplome de doctorat d'état option: géographie physique université mentouri Constantine.*

CL. Nesson, DJ. Sari, P. Pellon. 1978. *Recherche sur l'Algerie*. Edition du Centre National De La Recherche Scientifique 15, quai Anatole – France – 75700 Paris.

M.C. nesson, J. Valet, M. Rouvillois Brigol. 1973. *Oasis du Sahara algérien*, Institut géographique national – Paris – 1973.

Fraçois Paix. 1956. *Les nappes artésiennes de l'oued r'hir imprimrie baconnier alger 1956* in G. Toutain 1977 Eléments d'agronomie saharienne de la recherche – au – dévloppement Marrakech le 16 janvier 1977.

Michel Alain Roche. 1973. *Hydrogeologie de la haute saoura (Sahara nord occidental). Publications du service geologique de L'algerie* – Alger – 1973.

Benazzouz Mohamed Tahar. 2000. *Morphogenèse éolienne holocène et actuelle dans l'atlas saharien (Algerie): conséquences sur la désertification thèse présentée pour thèse présentée pour l'obtention du dilplome de doctorat d'état en géomorphologie universités mentouri constantine.*

Benazzouz M. Tahar. 2001. *Sebkhas et endoréisme dans les hautes plaines algériennes: cas des systémes endoréiques de l'est algérien.* Séminaire sur la valoristion des milieux semi - arides université de oum el bouaghi 28–29 mai 2001

Tesco, Viziterv. 1989. *Etude du réménagement et de l'extension des palmeraies de Oued R'high* , Synthése de l'étude, in Plan d'aménagement intercommnal 5 éme rapport Budapest. avril 1989.

Andre Voisin. 1964. *Le souf monographie 1954–1964.*

GENERAL PRESENTATION OF THE AL JERID AREA IN TOZEUR, TUNISIA, AS TEST AREA IN THE ARID LANDSCAPES FOR THE FIGHT AGAINST DESERTIFICATION

MOHAMED TALBI[1], ALBERTO MARINI[2], AND NABIL GASMI[1]
[1]*Institut des Régions Arides. Medenine. 4119. Tunisie*
Mohamed.Talbi@ira.rnrt.tn
nabil.gasmi@yahoo.fr
[2]*Earth Sciences Department – TeleGIS Laboratory – University of Cagliari*
marini@unica.it

Abstract

Desertification is the degradation of land in arid, semi-arid and dry sub-humid areas. It is caused primarily by human activities and climatic variations.

Over 250 million people are directly affected by desertification, and about one billion people in over 100 countries are at risk. These people include many of the world's poorest, most marginalized and politically weak citizens The international community has long recognized that desertification is a major economic, social and environmental problem of concern to many countries in all regions of the world.

Desertification can be reversed only if far reaching changes are made in local and international behavior. Step by step, these changes will ultimately lead to sustainable land use and food security for a growing world population. Combating desertification, then, is really just part of a much broader objective: the sustainable development of countries affected by drought and desertification.

In south western Tunisia, the legendary city of Tozeur, internationally well known for its Saharan oasis created thousands or year ago, as well as for its actual well growing Saharan touristic and cultural activities, is a good example of success story in combating desertification. Geographically speaking, the Tozeur region is located between two distinct geological fields – the Atlasic relief in the North and the Saharan platform in the South – the Tozeur area is characterized by structural aspects of transition. The synclinal undulations are not well expressed in the landscape; however, one can observe the development of the typical synclinal, which is intercalated between the mount of Drâa El jerid in the South and the dismantled mount of Ong el jemal in the North.

The averages of the high temperatures recorded at the station of Tozeur, coupled with long duration of sunlight, determine an enormous evaporating capacity of the climate. This high evaporation, combined with a weak rainfall, determines a strongly overdrawn annual balance. In the Tozeur region and despite these hard climatic conditions and the scarcity of natural resources, and thanks to its population who performed a participatory approach, the city succeeded to survive and to recreate a

A. Marini and M. Talbi (eds.), *Desertification and Risk Analysis Using High and Medium Resolution Satellite Data*,

favorable conditions for socio economical, cultural growth as well environmental protection and rehabilitation.

The following paper presents the historical, the human, the cultural aspects of the Tozeur area, and then highlights the climatic, the morphological and natural resources aspects and how the area managed to survive and develop a sustainable way of life for local population as well as for visitors.

Keywords: Arid landscapes, desertification combating, Tozeur, Tunisia

Introduction: The Jerid trough history are mainly grounds which lengthen from East to West, thus separating the two closed depressions *Chott El Jerid* in the South and *Chott El Gharsa* in the North. By this provision, the area of *Jerid* corresponds to a true isthmus, which allows the passage of Sahara to the South, with the hilly steppe sectors of the North. Within this framework, *Tozeur* at all times constituted an inevitable point of passage of the routes connecting these two sectors (Fig. 1).

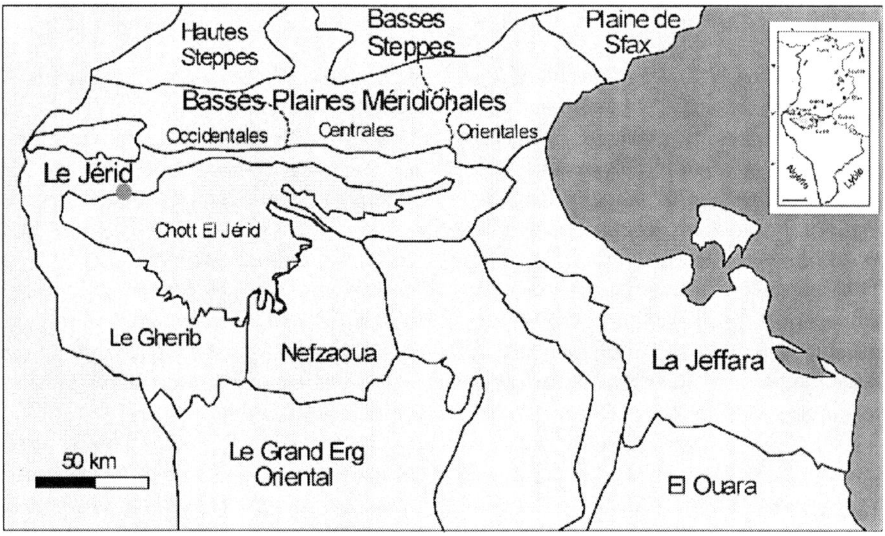

Fig. 1: The Jerid area, Tunisia

In addition, the presence of many sources of fresh water in such a hostile environment, allowed the concentration and the sedentarisation of populations since the ancient times. During these periods, the area was very attended by the caravans, coming to supply themselves of salt and/or to fill their requirement of water before crossing the Sahara (Fig. 2).

The blooming of civilizations of *Jerid* was marked by the development of the clever system of irrigation (Fig. 3) and the valorization of the rich agricultural knowledge accumulated during centuries.

This civilization, based on the valorization of water resources, made it possible for the *Jerid* population to offer to trade caravaneer new food products considered at that time as strategic.

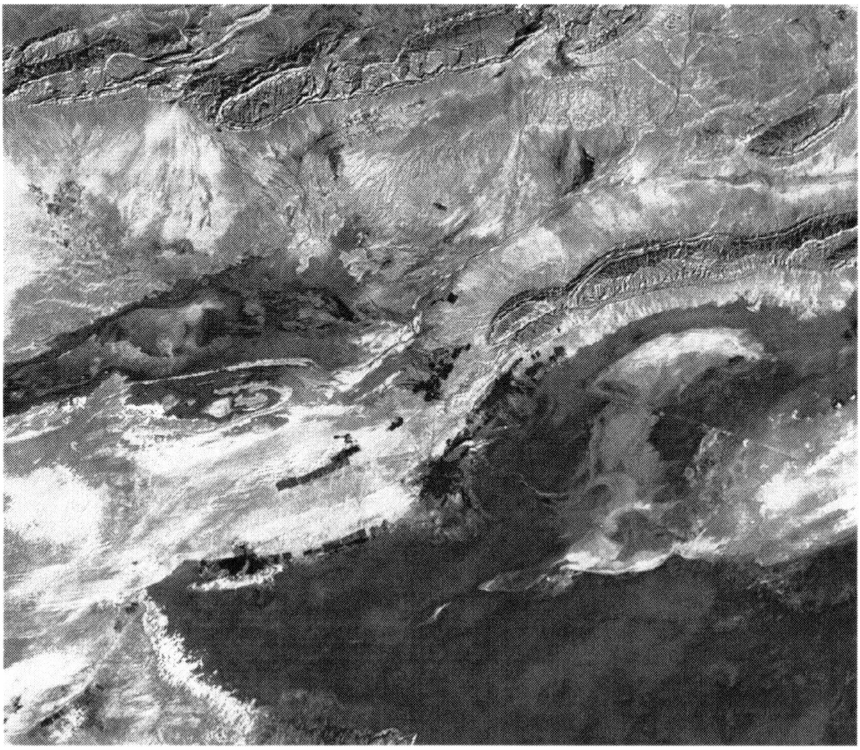

Fig. 2: Jerid indicates the whole of the Oases located at the close surroundings to the town of Tozeur. However, Jerid name nowadays covers a whole natural area of the south-west of Tunisia, located between Nefzaoua, Gherib and the low southernmost plains

Fig. 3: The distributor of the Charchara Wadi built with "ancient" blocks limestones

Thus, the abundant production of the multiple varieties of dates and products of the ground, also contributed at the role of this area in the commercial exchanges, on the local and regional scales, as well as at the supranational level.

The strategic position of *Jerid* led the Romans since the time of *Trajan* (reign between 98–117) to install a military station and the creation of colonies of veterans. The installation of military stations in various strategic points, thus constituting elements of the « *limes* » (Fig. 4), aimed on the one hand at extending the zone of Roman domination towards the South and on the other hand to protect the hilly areas against the invasions from the seminomads of the South and the nomads of the Sahara.

Later, during the Muslim time, the area of *Tozeur* maintained, if not reinforced, its position of privileged passage. It was reproduced thus on the main roads of the caravans of pilgrims "*er Rakb*", who started from "*Sakiat el Hamra*", on the Atlantic, towards their final destination, *La Mecque*.

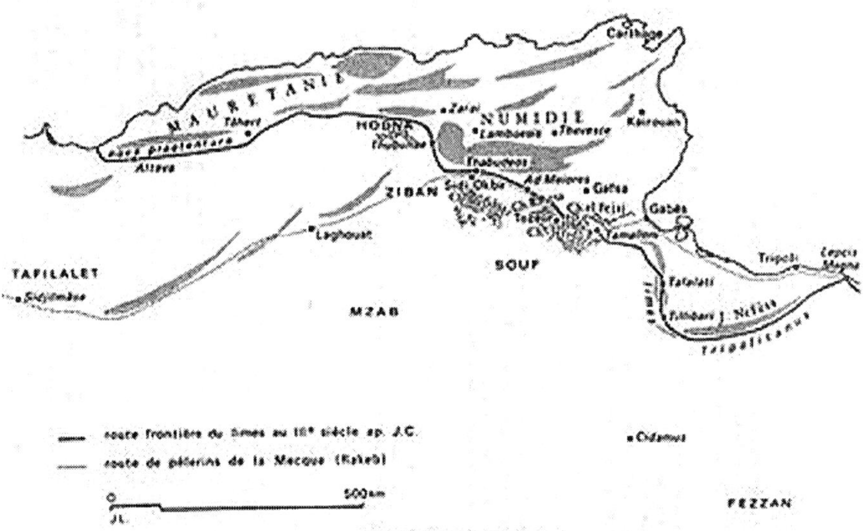

Fig. 4: Tozeur, principal relay of "limes" and "rakb"

The passage of the pilgrims constituted a special occasions of cultural and commercial exchanges and one could see the building of and the thriving of the multiples "*Zawya*" (place of worship and social exchange), left again in the area of Tozeur.

With French protectorate, at the end of the 19th century, the *chotts* algéro-Tunisian were in the center of a sharp scientific polemic and historical geography in France according to the assumption put forth with the whole beginning of the 19th century by English major *James Rennell.* This assumption, formulated in a work entitled *The Geographical System of Herodotus*, published in 1800, postulated the idea of the existence, at the 5th century before J.C., of an old bay largely open on the sea to the site of these large chotts.

Always according to *James Rennell*, this bay would have been gradually closed and its water would have dropped to give rise to the current chotts. This idea was included in a thesis supported in 1863 (*de tritonide lacu*) by the French diplomat and archaeologist *Charles Tissot* who, on the basis of the interpretation of certain old texts (Hérodote, the tour of Scylax at the 4th century before J.C., Pomponius Mela, 1st century of our era, Ptolémée, 2nd century) and of the doubtful indication of a discovered ancient ship within Nefta at the end of the 18th century, recognized in the *Chott Jerid*, the lake Triton.

This information was enough to constitute a lobby led by *Ferdinand de Lesseps* and including among its members many scientists, soldiers, politicians, business men, journalists and even men of letters as Victor Hugo who declared "*soldiers, astonish the universe by large things which are not wars*".

In addition, *Jules Verne* devoted to this project his last novel entitled *The invasion of the sea* (1905). Carried out by the commander François-Elie Roudaire (1836–1885), this lobby worked out a titanic, known project in the scientific literature under the name of project of *inland sea* in Algeria and Tunisia.

The project consisted in creating a water level of 8,000 km^2 (instead of 15,000 km^2 initially estimated and of which 6,700 km^2 are in Algeria), from which the maximum depth was to reach 31 m and this, through the digging of a channel of 200 km between the *gulf of Gabès* and the *Chott el Gharsa*, as well as the boring of three thresholds: the first with *Ouedref* (between the Mediterranean and the *Chott el Fjej*), the second with *Kriz* (between the *chott-el-Jerid* and the *Chott el Gharsa*) and finally the threshold of *Aslouj* (between the *Chott el Gharsa* and the *Chott Melghigh* in Algeria).

In parallel, one saw the constitution of a lobby, which set up against the project of *inland sea.* Thus, the geologist *Auguste Pomel* considered that such a project can "*speak with the eye but much less with the spirit*". The botanist Ernest Cosson saw in this project "*a scientific novel*". These scientists disputed the existence of an old connection of the chotts with the sea and asserted that the consequences of such a project were likely to be disastrous for the oases. The fights for influence and the diplomatic quarrels between the European colonial powers (France, England and Italy), ended up being terminating this project, which was abandoned in 1892.

Nowadays, the area of Tozeur always preserves its role of principal producer of best dates and particularly the variety "*Deglet Nour*". The characteristic of its oasis landscape, still radiant, the development of its potential of reception and the diversification of its centers of attractions contributed to the development of tourism, in particular Saharan.

The population of the area of *Tozeur* was 97,526 people in 2004. It was distributed as follows:

Urban population: 68,562, is 70.3%
Rural population: 28,964, is 29.7%
Source: the general census of the population and the habitat of 2004.

1 Morphological framework

1.1 Topographic aspects

Not very broken, modelled of surface of the area of *Tozeur* is appeared as three essential orographical units:

Low hills:

It is a relief far from being high. Its maximum altitude reaches only 185 m on the level of Drâa El Jerid. One announces approximately the lowering of the watershed in direction of the Western sector (altitudes of approximately 76 m). This shape compared to the neighbourhoods combines a configuration well lengthened from East to West on more or less 50 km. Its summit part draws a relatively arched form, while the slopes are not much contrasted in the landscape.

Plains:

The monotony of the landscapes of the area of *Tozeur* is caused by the extension of immense plane surfaces. They are indeed true plains, located mainly in the aureoles of the closed depressions. The extension of these plains develops more and more in the direction of the Western sector. Altitudes measured here, oscillate between 1 and 20 m. The slope of surface is directed mainly towards the closed depressions.

Chotts:

They are actually *sebkhas* and not chotts, because this last term indicates the aureoles of these depressions and present of the surface horizons of grounds rich in chlorides and sulphates and from which vegetable cover is characterized by predominance of halophytes species.

The area of *Tozeur* partially includes two closed depressions. Located at the north of Tozeur, *Chott El Gharsa* constitutes a true closed depression, which spreads out over nearly 1,350 km.

Stretched approximately 50 km and having altitudes lower than the sea level (from –5 to –25 m), this depression represents the basic level of the whole of the hydrographic network that drains the chains of the Southern Atlas and the Western part of the low Southern plains.

The name of *Chott* would be of Berber origin and would come from the word "*Aggersif*", resulting in: junction of river. This *Chott* constitutes, indeed, the natural discharge system of the wadis draining the slopes of the bordering reliefs (the *jbel Bliji* in North and *Drâa of Jerid* in the East).

The Southern part of the area of *Tozeur* consists in the development of a second closed depression. It is *Chott El Jerid*, which extends on approximately 5,400 m. Altitudes of this depression are higher than the sea level and oscillate between 15 and 20 m. This depression communicates towards the east with a third depression, called *Chott El Fejej*, covering a surface of 800 km.

The *Chott el Jerid* belongs to these spaces whose name changed on several occasions and who knew many peregrinations according to interpretations of ancient historiography.

- In the 5th century B.C., the Greek historian *Hérodote* was the first to evoke the *bay of Triton* and in that place ruined Jason and his Argonautes.

- In the 1st century of our era, *Pomponius Mela* located the *palus Tritonis* inside, behind of small Syrte (current *gulf of Gabès*) and without connection with the sea.
- In the following century, the Greek geographer Claude *Ptolomée* spoke about the existence of several lakes.
- In the 5th century, the priest and Spanish apologist *Paul Orose* placed *Lacus salinarum* at the east of Byzacène and Numidie.
- In the 11th century, *Abou Obeïd Al Bekri* called it *the large sebkha*, while *Abu Abdallah Tijani*, indicated in its *Rihla* (tour) beginning of the 14th century, the *Chott el Jerid* under the name of *sebkha Takamart.*
- In 1662, *Al Aïachi* crossed *sebkha-el-Kébira* (*the large sebkha*) and in 1709–1710, on his road for Mecque, the Moroccan traveller Moulay Ahmed crosses the *sebkha At-Tâkerma.*
- Finally at the 18th century, the Spanish monk Ximénès recognized the lakes quoted by Ptolomée in the Algero-Tunisian chotts.

It is finally useful to remark that the chotts of the area of *Jerid* as a whole constitute the greatest system of closed depressions of the North of Africa.

2 Structural aspects

Located between two distinct geological fields – the Atlasic relief in the North and the Saharan platform in the South – the area of *Tozeur* is characterized by structural aspects of transition.

2.1 Outcrops

The outcrops are only attributed to *Neocene*, from which one distinguishes two well contrasted levels:

2.1.1 The formation Beglia (Miocène)

The outcrops allotted to this formation wrap the majority of the mountainous links and apply to large surfaces. Their thickness varies between 50 and 400 m. It is likely that the dimension this thickness progresses in the West direction. These outcrops are characterized by a contrast in the facies with the former deposits (outcrop phosphatic of the Eocene). The terms fluvio-deltaic of the *Beglia formation* (Miocène) covers in discordance the *Eocene* and *Cretaceous* layers of marine origin. The facies of this formation is characterized by the following succession:

- The contact with the subjacent outcrops (cretaceous, Eocene or, by places, lower miocene) is materialized by the individualization of a not very thick horizon (a few centimetres) constituted by conglomeratic elements.
- The average part is much thicker (approximately 2/3 the total thickness of the formation). It consists of an alternation of relatively fine sandy terms with stratification intersected with green and decimetre gypseous clay beds.
- The upper part is characterized by rarefaction of the gypseous beds of clay and predominance from white sands yellowish and rich in dragees from silica.

2.1.2 The formation Ségui (Pliocène)

It is about the higher lithological element of the "continental complex". It constitutes a thick layer (150–500 m) which wraps almost the totality of the outcrops of *Beglia* (*Miocène*). The lithology of this outcrop is characterized by the following succession:

- At the base, a conglomeratic bench of a few meters thickness.
- At the top of the first conglomeratic bench, a thick layer (almost half the total thickness) accumulates, made up of relatively consolidated red clays. One also notes the gypsum omnipresence which appears mainly by nodules and locally clusters with farinaceous structure.

The higher term is marked by the succession of the conglomeratic benches. This continental deposit is dominated by the coarse elements (blocks and stones limestones) resulting from the mountainous systems. It is thus the most resistant term of the Neocene outcrops. This consolidation is supported by the development of a thick gypseous crust on the surface.

2.2 Structural morphology

Extended approximately in the orientation of the Northern chain of the *chotts*, the area of Tozeur preserves the witnesses of a crumpling Atlasic, but of minor scale (Fig. 5).

2.2.1 Mountains

The mount of *Drâa El Jerid* records a considerable axial lowering in opposition to the whole of the mountains of the Northern Chain of the chotts. It is possible that this lowering is in connection with the development of a network of directional faults (is/western), which affects the Northern and Southern side. In addition, other anticlinal undulations of minor scale appear in the Southern part of *Chott El Gharsa* and in the North of the village of *Hazoua*. These convexities are represented almost as tender lithological units, showing models relatively eroded.

2.2.2 Hills

The synclinal undulations are not well expressed in the landscape, however, one can observe the development of the typical synclinal, which is intercalated between the mount of Drâa *El jerid* in the South and the dismantled mount of *Ong el jemal* in the North. It is about a not very wide valley and relatively covered by wind accumulations.

3 Climatic characteristics

According to the index of Emberger, the area of Tozeur belongs on the *higher Saharan bioclimatic level*. Its climate is ordered primarily by three factors:

- The continentality of the zone and its distance with respect to the maritime influences, implying very high diurnal and annual thermal variations
- The position on the Northern edge of the Sahara opening the area with the influences of the subtropical high pressures
- The mountainous chain of the Southern Atlas which constitutes a barrier vis-à-vis the influences of the North

Fig. 5: The "Mihrab" of the Large Mosque, example of Muslim medieval architecture

3.1 Temperatures

The very marked continentality of the area of Tozeur is expressed by a very contrasted thermal mode with very marked annual and diurnal amplitudes. The average temperature is about 22°C. In fact, its climate is marked by two seasons (Fig. 6): The *hot season,* starts in May with temperatures higher than 22°C and lasts 4 to 5 months (May–September). The average temperature during the hot season is in the neighbourhoods of 30.5°C, whereas the maximum ones easily exceed 50°C.

The *fresh season*, is characterized by soft winter, with monthly average temperatures always higher than 10°C. But the minima of the coldest month can go down to less than 4°C, with a considerable fall of the temperatures during the night, accompanied sometimes by frosts (Fig. 7).

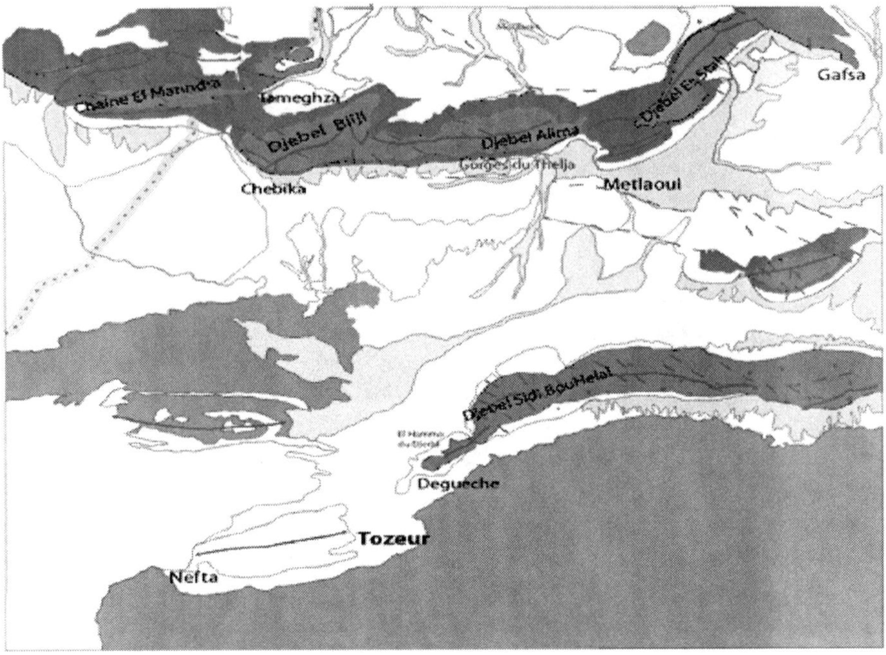

Fig. 6: Chart of the cretaceous and Neocene geological outcrops (INM/Tunisia)

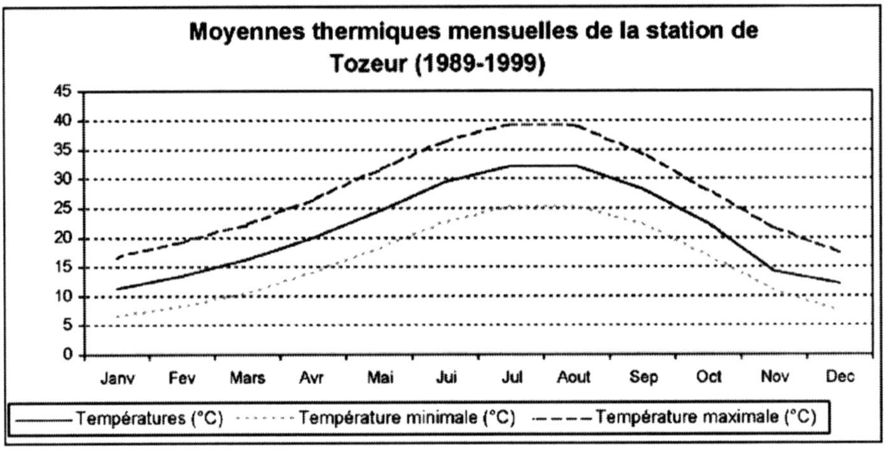

Fig. 7: Tozeur mean monthly temperature between 1989 & 1999 (After INM/Tunisia)

3.2 Pluviometry

The area of *Tozeur* is located in the lower part of the isohyet 100 mm and often lower than 50 mm. Precipitations are rare and irregular. The average of rainy days is of 20 days/year. Minimum being of 6 days/year and the maximum of 28 days/year.

The illustration below (Fig. 8) illustrates the pluviometric mode at the station weather of Tozeur for the period going from 1900 to 1995.

The monthly distribution of precipitations (Figs. 9 and 10), reveals March and November (or sometimes September) as being the most sprinkled months. According

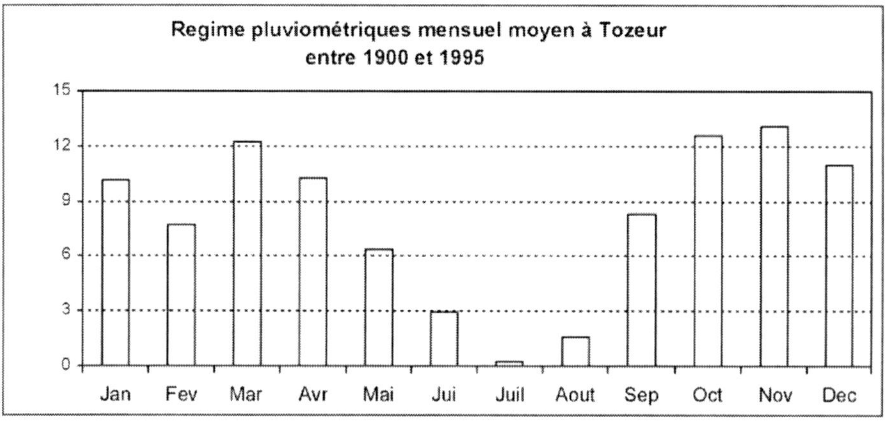

Fig. 8: Tozeur mean monthly rainfall between 1900 & 1995 (After INM/Tunisia)

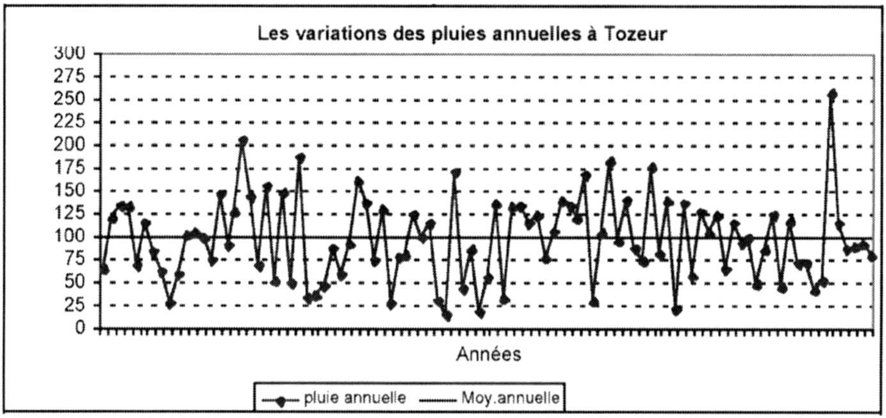

Fig. 9: Tozeur yearly rainfall variation to mean, between 1900 & 1995 (After INM/Tunisia)

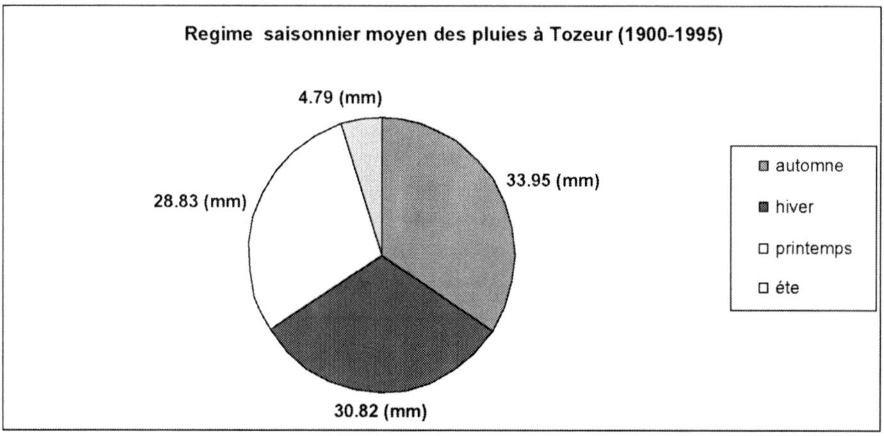

Fig. 10: Tozeur seasonal rainfall repartition between 1900 & 1995 (After INM/Tunisia)

to the seasons, Autumn is famous for being the rainiest season. Indeed, the rain season begins more or less late (often at the 3rd decade of September), whereas its end is early (3rd decade of April).

It proves to be interesting to announce, in this context, that the return of the wettest years occurs about every 20 years.

3.3 The water balance

The averages of the high temperatures recorded at the station of Tozeur, coupled with long duration of sunning (Fig. 11), determine an enormous evaporating capacity of the climate. This high evaporation, combined with a weak pluviometry, determines a strongly overdrawn annual balance (Fig. 12).

The permanent deficit is from about 1,400–1,800 mm/year and increases in direction of the *Eastern Erg*. Thus, the received rain cover only 10–15% of the ETP. In fact, the hydrous deficit, which is permanent in the zone of study, is strongly constraining, as well for the natural vegetation as cultivated.

3.4 Wind

The winds play a central role in the morphological dynamics of the arid regions. The zone of study is subjected to important flows of wind drained by a vast network of *wind corridors* through the assembly lines bordering the chotts, and laid out according to a west-east axis (chain of *Jbel Negueda-Jbel Enneflet*, chain of *jbel Sidi Bouhlel, Jbel Tebaga and Fatnassa,...*), like through the discharge system of *Biskra* (wind of west) and while passing by the *Eastern Erg*.

Thus, one records in *Tozeur* 120 days/year strong wind, with a clear predominance of the winds in NE direction in autumn and in spring. While in the summer, the winds of direction SE dominate.

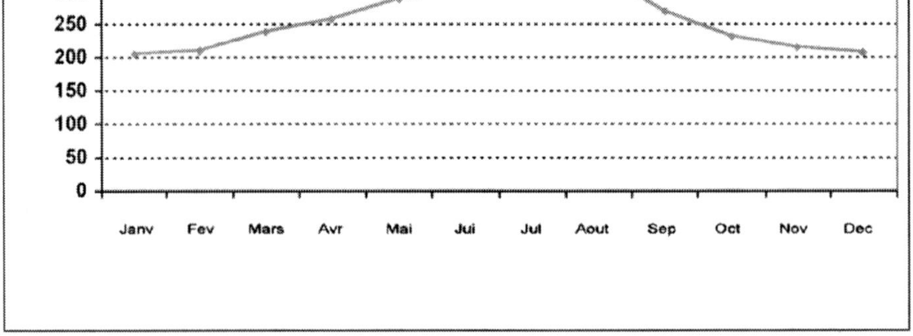

Fig. 11: Tozeur mean monthly sunlight between 1981 & 1995 (After INM/Tunisia)

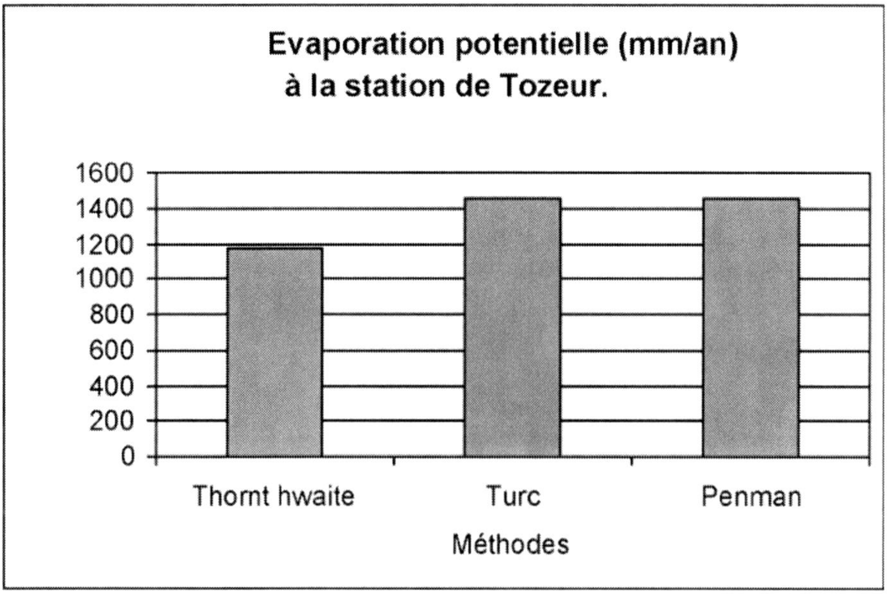

Fig. 12: Tozeur mm/year evapotranspiration (After INM/Tunisia)

The rose of the winds (Fig. 13) shows that the winter is marked by the frequency of the winds from West direction. The significant proportion of the winds which blow from the western sector during the winter is similar to the circulation of the West, which dominates almost the totality of Tunisia, particularly during this period.

The *Tozeur* area is characterized by the frequency of sand winds. The average is there of 74 days/year of sand wind, whose 45.9% take place in summer. The frequency of the days of sand increases gradually in the Western zone, in direction of the Great Eastern Erg.

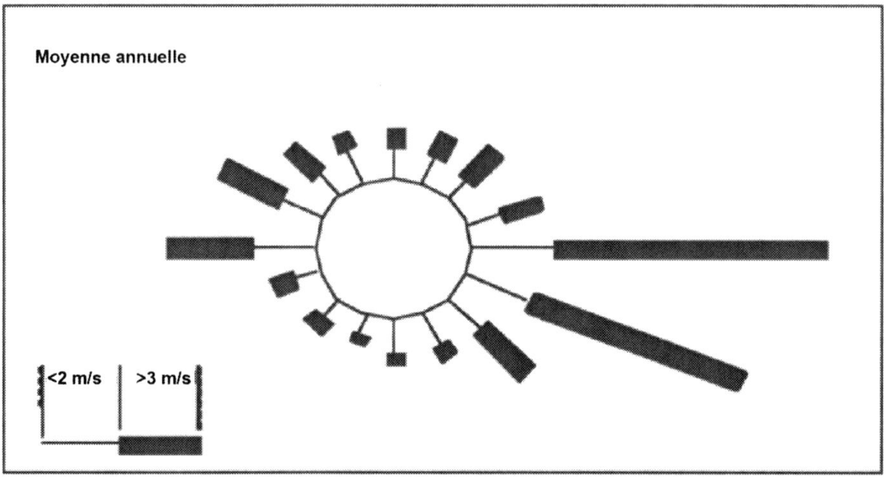

Fig. 13: Winds distribution in Tozeur (period 1989–1999) (National Institute of Meteorology)

This type of wind has fatal consequences on the cultivation, in particular during the summer, considering that it coincides with hot and dry winds (Sirocco) and an abrupt rise of temperature (of 10 to 15°C in 1 or 2 h).

Let us finally announce that the winds constitute a source first order permanent natural harmful effect, and therefore it is necessary to hold account in all the actions of adjustment of territory and economic development.

It is in this context that the projects of development focus much on the large interventions of fight against the desertification, which according to these organizations, should contribute to reduce the abrasive action in particular around the oases.

4 Natural resources

4.1 Soils resources

According to the available data, and in particular the pedological chart of Tunisia (scale 1/500,000), the area of study is characterized by soils not very wide and very localised. The scarcity of the soils in this area finds its explanation in the mentioned aridity, a vegetable cover very scarce, if not absent, in addition of the depressions closed during the Quaternary to the influence on the quality of the soil.

By adopting a transect from top to bottom one can distinguish four types of soil:

4.1.1 Halomorphic soils

They characterize the Northern and Southern parts of the area of study. They rest on alluvial materials with sablo-muddy texture with limono-sand spreader. They present, especially during the dry season, the shapes of efflorescence of salts due to the increase of the ground water very salted and loaded with gypsum. At the end of the dry season, this clay and salt film disaggregates under the effect of dryness. The result is the formation of a surface horizon of a few centimetres thickness (pseudo sands).

The halomorphic soils show by places (the zones of beat of the ground water, the circumference of *Chott El Gharsa*, the downstream part of the old oases), the presence of very varied gypseous accumulations (cluster, nodules, crusts and encrustings, roses of sand, microcrystals, terch, deb.-deb....). Generally, these grounds are not cultivable, except for the parts occupied by the oases, conditioned by the intensive irrigation and an organic matter and sand contribution, which leads to their scrubbing.

4.1.2 Soils of the oases

They are anthropic soils, gained at the expense of the halomorphic soils. Indeed, with the intense irrigation during one long period, the halomorphie decreases and is shifted on a lower level. Organic matters and detrital elements (sand) contributed with time to modify the halomorphic soils at all the levels (structure, fertility, texture...). Thus, it is frequent to observe in the traditional oases a horizon of 40 cm thickness, rich in organic matter.

The soils of the oases are however very sensitive to the salinisation and require an effective system of drainage. In the places where one frequently uses salty water for the irrigation, one can see a clogging and a concentration of salt on the surface.

Lastly, a distinction is to operate between the soils of the traditional oases and those of the recent oases: the latter developed obligatorily near the water points, which gives place to soils more appropriate to the cultivation, at least during the first years of introduction.

4.1.3 Gypseous soils

In the zone of Tozeur, the outcrop of these soils occurs at the edge of *Chott El Gharsa* and in the zones of extension of the gypseous crusts or the gypseous geological formations. It acts, indeed, of the grounds rich in gypsum (rate of 70%), little or not cultivable and carrying only thin natural vegetation, being used as pasture in particular with camélidés.

4.1.4 Rough mineral soils (blow sands)

The zone of transition between *Chott El Jerid* and the *Eastern Erg* is marked by the presence of dunes of various forms and sizes. The proximity of the *Eastern Erg* partly supported the extension of these forms.

Given their mobility, these dune accumulations do not present a pedogenic horizon. They are in fact rough mineral grounds not very suited to the cultivation.

Generally, except for the artificial grounds of the oases, the area of study presents only skeletal soils distinguished by a reduced depth, a very fragile structure and a low content of organic matter, inherent in particular with the weakness of vegetable cover. In the case of the grounds of the irrigated perimeters, the abundance of the ions Na+ (sodium) in water leads after irrigation and – fault of an effective system of drainage – to situations of asphyxiation, very unfavourable to the plants. All these characteristics generally lead to decrease the water holding capacity of the plants. Thus, the reserve usable in this type of ground is evaluated with 0.5 mm/cm.

4.2 Water resources

The water resources take a very particular importance in any development project in the arid areas, water being "the source of the life".

In addition, the zone of study is distinguished by the almost total absence from surface water. This situation is justified by the insufficiency of precipitations, combined with the influence of evaporation.

So the water resources are limited to the ground water and deep.

4.2.1 Ground water

The zone of study is occupied by the phreatic surface of Jerid, said water table of oasis. Its alimentation primarily coming from the "drainance" from the underlaying phreatic surface and the surpluses of irrigation water. Thus, these exploitable resources are evaluated to 14.3 million m^3/year.

In spite of the poor quality of water (Rs: 4 with 6 g/l), this water table currently knows an overexploitation in the area of Jerid. The extraction carried out is about 23 million m^3/year, with a tendency to increase, considering the requirements out of water increasing for the oases. In fact, the water table of Jerid presents an overdrawn assessment of 8.7 million m^3 and knows a noticeable decrease of its piezometric level, which reduces water quality.

4.2.2 Deep groundwater table

The zone of study is covered by two following aquiferous systems:

- *Groundwater table of the Final Complex* (*complex terminal – CT*)

It constitutes the main aquifer of all the area of *Jerid-Nefzaoua* considering its vast wide, energy of southern bank of *Chott El Gharsa* in extreme cases Western of *Dahar*. In the zone of *Jerid*, the aquifer is of sandy nature, which supports a slow current food starting from precipitations. Water of this phreatic surface is essentially fossil, therefore nonrenewable, and the underground flow is carried out from south towards north in direction of *Chott El Jerid*. The destocking is estimated at 10 m^3/s. The phreatic surface of the *final complex (complex terminal)* was marked in *Jerid* by the gushing of the sources, thus favoring an artesian system in all the area. On the other hand, the salinity of water reaches values rather high in the area (Rs: 2.5 with 3.5g/l with *Hazoua and Nafta*). Water of this phreatic surface is exploited for 90.8% in the agricultural sector, while the drinking water accounts for only 3.1%. The phreatic surface "*complex terminal*" currently knows an overexploitation in the zone of *Nefzaoua*. Whereas in the sector of *Jerid*, the mobilization of the resources of this water table did not reach the critical point.

- *The water table of « Continental Intercalaire » (CI)*

It covers all the Tunisian-Algerian Saharan border zone. Also called "*nappe albienne*", it corresponds to a multi-layer water tables developed in the detrital series (sand or clay) of the lower cretaceous, and located at a depth from 1,300 to 2,500 m. Moreover, the tablecloth of the Continental Guide is essentially fossil. Its current alimentation starting from *Dahar* is very limited.

In *Jerid*, water of the *Continental Intercalaire* is characterized by their high temperature (between 70°C and 76°C), combined with a strong salinity, about 5 g/l, which makes them unusable for drinking water. In spite of the problems that it pose (investment for the installation of equipment of cooling…), the warm water constitutes an additional potential, considering their importance in the installation of the green house perimeters with market-gardening vocation, within the framework of the "National strategy of Use of Geothermal Water in Agriculture" (SNVEGA). In *Jerid*, the exploitation of the *Continental Intercalaire* is done by major drillings. Moreover, the *Continental Intercalaire* is exploited currently only with proportion of 15% (7.5 million m^3/year).

In the zones of *Hazoua and Nafta*, water quality becomes poor, therefore this freatic table is used nearly completely for agriculture. But its use for the irrigation is conditioned absolutely by an effective system of drainage.

The strategic plans for the exploitation of the water resources of the Tunisian South, as well as complementary studies, allowed in this respect to fix, since 1994, the potentialities of the deep water table (Fig. 14).

In addition, the requirements out of water for *Jerid* in year 2000 were limited to abot 190 million m^3/year.

Deep water table	Potentiality in the year 2000	Productivity (m^3/year)					
		1981	1985	1990	1995	2000	2010
« Complexe Terminal »	151.37	102.12	121.63	143.42	150.23	157.27	157.27
«Continental Intercalaire »	63.07	0	8.83	26.49	31.53	31.53	31.53

Fig. 14: Potentialities of the deep phreatic surface envisaged on these waters from 1981 to 2010

It proves to be interesting to recall that the multiplication of drillings accelerates the fall of the artesianism, the recourse to pumping, the increase in the cost of water; and so the amplification of the problem of hydromorphism in the oases.

4.3 Vegetation

The area of study is dominated by a pseudo steppe with psammophile species and gypsophiles, characterized by three levels (from 50 to 200 cm, from 15 to 50 cm and from 0 to 20 cm). This steppe ensures a covering of the ground of about 20% in the best situations. The pasture offered is as a whole poor and constitutes only one summer supplement in particular for camélidés.

The primitive vegetation is Tamarix. But one can generally distinguish three vegetable groupings, in close connection with the type of the ground:

- Grouping with *Haloconemum Strobiloceum*, *Arthrocnemum indicum*, *Salsola Siberico*, *Salsola tetrandra*, *Sueda mollis* on the salted formations
- Grouping with *Aristida pungens* (*Stipagrostis pungens*), *Retama raetam*, *Calligomum* on the sandy formations
- Grouping with *Astragalus armatus*, *Zygophyllum album*, *Erodium glaucophyllum* on the gypseous formations

5 Conclusion

Located at the septentrional border of *the Great Eastern Erg*, the zone of study is defined as a hyper-arid space with weak and irregular precipitations. The winds are strong and dessicants, the hydrous assessment is largely negative.

The area, at the same time, is presented as a fragile and hostile medium. The brittleness of the natural environment and the risks of erosion of the ground are reinforced by a precarious ecosystem:

- *Skeletal and little extended soils*
- *A rare and very degraded natural vegetation, if not absent*
- *Water resources of very poor quality, with a main risk of exhaustion*

Taking into account the increasing scarcity of the water resources, combined with a higher and higher increase in the intersector needs, as well on the quantitative level as qualitative, it is necessary to rationalize the exploitation and the management of the water resources in term:

- *Of control of the extension of the private wells and consequently of the overexploitation of the tablecloths*
- *Of improvement of the delivery system and management of water of irrigation in the oases*
- *Of improvement of the techniques of irrigation and this by the generalization of the systems of irrigation for saving water*
- *Of sensitizing of the population and engineering services with the risks of exhaustion of the water table and the means of prevention*
- *Of development and diversification of the less consuming water cultures*
- *Of orientation of the programmes of use of waste water of the agglomerations and oases in agriculture: fodder, windscreen, industrial crops...*
- *Of encouragement of the valorization of water of drainage around the oases for the irrigation of the trees and multi-purpose shrubs (windscreen and fodder on feet)*

In this context, the Decennial Strategy of Water resources (SDRE), which covers the period between 2002 to 2013, considers that the planning and the management of the water resources are a vital need to ensure the water safety of the area.

Bibliography

Anversa L., Bentivegna V., Ferretti L.V., Filpa A., Lacava A. 1990: *Tunisia: rehabilitating oasis towns.* Quaderni del Centro Analisi Sociale, 142 p.

Bagnold R.A. 1953: *Formes des dunes de sable et régime des vents.* Colloque international, c.n.r.s., pp. 23–32.

Ben Ouezdou H., Gasmi N., Boussida L. 1996: *Contribution à l'étude géomorphologique du Jbal Sidi Bouhlel et ses environs (sud-ouest de la Tunisie).* Revue Tunisienne de Géographie, No. 30, pp. 53–80.

Blum M., Carbondele I., Swezy C.S., Kocurek G., Deynoux M., Ben Ouezdou H., Gasmi N., Bouossida L. 1996: *Contribution à l'étude géomorphologique du Jbal Sidi Bouhlel et ses environs (sud-ouest de la Tunisie).* Revue Tunisienne de Géographie, No. 30, pp. 53–80.

Bousnina A. 1986: *La variabilité des pluies en Tunisie.* Pub. Université de Tunis, série geogr, Vol. 22, 308 p.

Count Byron Khun de Prorok 1925: *Ancient Trade Routes from Carthage into the Sahara.* Geographical Review, Vol. 15, No. 2, pp. 190–205, American Geographical Society.

Drake N.A. 1997: *Recent Aeolian Origin of Surficial Gypsum Crusts in Southern Tunisia: Geomorphological, Archaeological and Remote Sensing Evidence.* Earth Surface Processes and Landforms, Vol. 22, No. 7, pp. 641–656, Wiley, Chichester.

Khatteli H. 1996: *Erosion éolienne en Tunisie aride et désertique: Analyse des processus et recherches des moyens de lutte.* Thèse de Doctorat Es-Sciences biologiques Appliquées. Univ. de Gent, 180 p.

Le Houerou H.N. 1995: *Considérations biogéographiques sur les steppes arides du Nord de l'Afrique.* Rev. Sécheresse, Vol. 6, No. 2, pp. 167–182.

Moncef Chalbaouia and Hamed Ben Dhia 2004: *Main characteristics of reservoirs in the southwest of Tunisia.* Comptes Rendus Geosciences, Vol. 336, No. 13, pp. 1191–1198.

Perotti C., Rinaldi M., Marini A., Muntoni F., e Zucca F. 2007: *Studio morfo-strutturale del fronte S dell'Atlante tunisino attraverso l'analisi di immagini stereoscopiche ASTER*. Rivista italiana di Telerilevamento, Vol. 39, pp. 117–134.

Santucci G. 2002: *Ai confini dell'Impero: contatti, scambi, conflitti*. Atti del XV Convegno di studio, Tozeur, 11–15 dicembre Africa Romana.

Talbi M., Gasmi N., Jguirim H., Ben Mansour N., Talbi K. 2000: *Savoir local et gestion optimale des ressources en environnement aride: Cas de la région de Sehib (basses plaines méridionales)*. Séminaire International MEDNPOP 2000. Jerba-Tunisie. pp. 1–8.

Trousset P. 1982: *Le franchissement des chotts du sud tunisien dans l'antiquité*. Antiquité africaine T18. P. 45–59.

MAPPING DESERTIFICATION, A HIGH EDUCATION COURSE IN REMOTE SENSING

ALBERTO MARINI[1] AND MOHAMED TALBI[2]
[1]*University of Cagliari – Italy CINSA Telegis Lab.*
marini@unica.it
[2]*IRA Insitut des Regions Arides – Medenine Tunisia*
mohamed.talbi@yahoo.fr

Abstract

The participants to this Course of high formation are researchers from Algeria, China, Egypt, Italy, France, Belgium, England, Morocco, Mauritania and Tunisia, guided by teachers from Algeria, Belgium, England, Egypt, Italy, France, and Tunisia.

The program has previewed conferences about the phenomena of the desertification, some guided visits, projects of fight against the desertification, projects about the conservation of natural resources. The programme included also practical guided work on the field, related to survey systems mainly based on images from satellite, in order to obtain maps to support the land planning.

The works have been followed by a general discussion that allowed to reach the following conclusions:

The desertification is an universal phenomenon, it exceeds the frontiers of the different countries; hundreds million of persons all over the world are directly touched;

- Scientific research has an important role and must be oriented toward the applicative aspects in a strict relation with the demands of development.
- A reliable information in constant modernization is a necessity and a guarantee for the long-lasting management of the natural resources.
- The techniques of survey, particularly remote sensing, as well as geomatic, constitute privileged ways in order to control the desertification and to support the necessary decisions.
- Indeed, these techniques allow to measure the impact of the phenomenon on the people and on the natural environment.
- An interdependent and international partnership has primary importance, particularly in the field of the inventory, and in the control and the evaluation of the land deterioration in barren zones. A general visit crossed the various landscapes units along a route of more than 200 km. This first contact with the land was indispensable in order to understand better several aspects of the barren environment.

The successive field surveys were based on a detailed cartography, whose objective was not limited to the level of the apprenticeship of the cartographic survey methods. It was also necessary to realize maps easily employable by the people in charge of the fight against the desertification.

A. Marini and M. Talbi (eds.), *Desertification and Risk Analysis Using High and Medium Resolution Satellite Data*,

In this context the participants have been subdivided in three groups of various nationalities and various disciplines.

Each group was responsible for a specific field of study. The thematics developed for the detail surveys have been:

- Taking advantage of the images from various satellites, with various ground resolutions (MODIS, 250 m, ASTER, 15 m, Quickbird, 0.6 m).
- Surveyed for land use maps and sketch maps about morphodynamics and field vulnerability, for the entire zone of study and each representative test area.
- Annotations and discussions have been organized for each phase, in order to draw the necessary conclusions, to understand the dynamics of the country and to realize the local thematic cartograph whose objective was not limited to the level of the apprenticeship of the cartographic survey methods.

Participants have developed the job in a homogenic group and with professional conscience, but also challenging the difficulties that they encountered (accessibility of the land, sandstorms, rains...).

After the surveys the jobs were displayed in the center and the results of each group was presented in front of a "jury" composed by teachers, the members of the other two groups and also local personalities and experts, and finally by specialists invited for this occasion.

Keywords: Building technical capacities, environmental change mapping, remote sensing, desertification combating, south-north partnership

1 Introduction

In Tozeur, with the North African partners, we have elaborated a research plan on the topic of the land use and, in particular, on the impacts of the environmental emergencies (as desertification, climate change, degradation) at the economic and social level. Such line of research can also be connected with partnership that the TeleGIS in Cagliari has in the Antenna Network, also with Eastern European Countries. Our analysis made in Touzer has focused on the importance of the environment of the oasis, strongly threatened from the emigration of the people who work in these not natural habitats.

The results of this possible research should be a reference also for areas with a different landscape.

1. For the closing of 2006, year dedicated to the Desertification, the TeleGIS Laboratory of the University of Cagliari (Italy) has organized in Tozeur (Tunisia) the formation course "Practical use of high and medium resolution sensors for the valuation and the monitoring of sectors in desertification or in productivity leak", with the funds received from the NATO and with the aid of the Institute of the Barren Regions of Medenine, that has celebrated the 30th year from its foundation and the renew of the quinquennial collaboration with the TeleGIS Laboratory.

2 Conferences and presentations

2.1 Conferences (scientific sessions 1 and 2): The first day has been dedicated to conferences about the general aspects of the desertification, to the various aspects of the barren landscapes and to their dynamics. A special session has allowed the participants to introduce their present and future works. The third session has been dedicated to the aspects of remote sensing and to the geographic information systems; to the multiple applications of these surveying systems as tools for the fight against the desertification and for the reasoned management of natural resources. Those three sessions have been followed by a general discussion that allowed to reach the following conclusions:

- Desertification is an universal phenomenon, it exceeds the frontiers of the different countries; hundreds million of persons all over the world are directly touched.
- Scientific research has an important role and must be oriented toward the applicative aspects in a strict relation with the demands of development.
- A reliable information in constant modernization is a necessity and a guarantee for the long-lasting management of the natural resources.
- The techniques of survey, particularly remote sensing, as well as geomatic, constitute privileged ways in order to control the desertification and to support the necessary decisions.
- Indeed, these techniques allow to measure the impact of the phenomenon on the people and on the natural environment.
- An interdependent and international partnership has primary importance, particularly in the field of the inventory, and in the control and the evaluation of the land deterioration in barren zones.

2.2 Some visits to a series of representative localities in the region have allowed the participants to better understand the negative impacts of the desertification on the people and on natural resources and, at the same time, allowed them to estimate the efforts undertaken from Tunisia in order to fight against the desertification.

These efforts are realized to the advantage of the population and the environment in the general actions of the national Program of fight against desertification (PAN-LCD), with an original approach, integrated with a long-lasting development.

2.2.1 During the visit to the Center of development of biodynamic agriculture (GDABD), installed in Hazoua, the participants have assisted to a presentation about the concepts and the techniques of this type of agriculture and about the different kinds of production.

In a guided visit to a connected portion of land, the participants have been able to verify the originality of the work realized, which represents a guide for differentiating the activity of the peasants to guarantee them a better rendering; this guarantees the duration of the investment and the valorization, with the consequence to guarantees the diminishing of pressure on the environmental resources.

2.2.2 Moreover, the visit carried out in the portion of land object of reforestation on the border of the modern oasis of Chamsa and Ibn Chabbat has allowed the participants to estimate the efforts undertaken in order to value salty waters that from the deep wells flow in excess from the oasis and are used for the irrigation of the forest plantations and for the pasture lands.

The participants have praised this initiative and have emphasized the effectiveness of the participation and the desire to see it spread in their respective countries.

2.2.3 The center of research on the dates palms, situated in Déguèche, has presented a relation on the guidelines of the research of the same center.

The participants have been able also to observe the various interventions carried out on national, maghrébine and international scale, they had the chance to visit the collection of palms of the center (more than 100 different varieties of dates palms).

2.2.4 At the end of these visits, the participants have appreciated the undertaken efforts and the results obtained.

They also learned from the Tunisian experience the following positive lessons, to be valorized in other hit countries:

- The desertification is a problem connected to the development.
- The fight against the desertification is to be integrated in the national, regional and local programs.
- The creation of alternative activities as complementary for the population, constitutes one of the main guarantees of the success.
- The protection of the resources and the use of the renewable energies need to be encouraged.

3 The survey and the results

Under the control of the experts and in the framework of the "practical formation", the participants had to elaborate a series of thematic maps inherent to the desertification, helpful as support for planning and taking decisions.

In order to obtain these aims a technical session concerning the methodology to use for the elaboration of the satellite images has been organized.

The participants received from the pool of experts a formation on:

- The interpretation of satellite images in the study region (panchromatic and false colour images)
- The preparation of the work of acknowledgment of the land and the identification of the routes for the survey campaigns
- The compilation of the reference data
- The argument and validation of the cartographic base

3.1 The first field survey has interested all the participants of the group. A general visit crossed the various landscape units along a route of more than 200 km: this first contact with the land was indispensable in order to understand better several aspects of the barren environment.

For this purpose, with the contribution of the experts, many detailed analysis have been elaborated. Annotations and discussions have been organized for each phase, in order to draw the necessary conclusions, to understand the dynamics of the country and to realize the local thematic cartography.

3.2 The second field survey was based on a detailed cartography, whose objective was not limited to the level of the apprenticeship of the cartographic survey methods.

It was also necessary to realize maps easily employable by the people in charge of the fight against the desertification. In this context the participants have been subdivided in three groups of various nationalities and various disciplines. Each group was responsible for a specific field of study.

The thematics developed for the detail surveys have been:

- Taking advantage of the images sent from various satellites, with various ground resolutions (MODIS, 250 m of resolution, ASTER, 15 m, Quickbird, 0.6 m)
- The elaborations, for the entire zone of study and each representative test area, of land use maps and thematic maps about morphodynamics and field vulnerability

4 Estimation

4.1 External appraisal of results

From the opinion both of the jury and of the local authority for the fight against the desertification the participants, despite the short period and the difficult climatic conditions, have realized a job of great practical usefulness for the local development.

In fact, the conclusions and the topics developed are an important base in order to strengthen the efforts undertaken, but also in order to develop the fight against the desertification according to an absolutely original approach and vision.

4.2 Self-evaluation of the formation by the participants

At the end of the seminar the participants have been invited to express their opinion on the organization, through the compilation of an evaluation form they gave personal appraisal on:

- The topic of the seminary, and its duration
- The conferences introduced
- The methodology used
- The usefulness of the visit to the land
- The different specializations of the participants and the use of many languages and their impact
- The strong points and the weak points
- The proposal of topics for future sessions of formation
- Finally, the usefulness of this formation for the professional life of the participants

4.2.1 From the analysis of the forms, turns out in particular that:

- 100% of the participants find the topic of the seminar interesting.
- 64% think that the duration of the seminar (10 days) is appropriate, while 21% find it long and 16% consider it short.

- Concerning the connections of the conferences with the seminar topics, approximately a third of the participants consider this correlation as strong. However, more than 2/3 of them (69%) find the correlation between the conferences that have been presented and the topic of the seminar only acceptable. This remark is important for possible future sessions (there was a significant difference in the field preparation of the participants).
- Concerning the usefulness of the visits to the field illustrating the theoretical aspects of the formation, 100% of the participants consider them as very profitable.
- Concerning the methodology used during the high level formation (to learn through research and responsibilization), 68.5% consider the approach original, while 31.5% consider it classic.
- Answering the question about the heterogeneity of the work groups (different specializations, various nationalities...), 79% find the experience as positive and very valuable and 21% find it normal.
- To the question aiming to assess if the multilingual approach has placed problems (Arabic, English, French, Italian during the formation, rather than English, as it was previewed or French as it would be preferred) 5% of the participants answered that it created some problems, 31.5% think that it did not and, on the contrary, the 63.5% consider that this experience has been for them much beneficial.
- As an answer to the question of the impact of the seminar on their professional life, 79% of the participants give positive answer and 21% think that the impact is medium.
- The weak points that have been remarked, and that it will be necessary to avoid in the future concern the limited information that the participants received before their arrival to Tozeur, the scarcity of the documentation; the lack of sessions of formation on the software.
- For the proposals of future formation, four topics are more impressive. They are:

  1. A deeper formation in cartography of environmental dynamics
  2. The management of natural risks
  3. The integrated development of barren zones
  4. The management of water resources

Concerning the last question related to a general evaluation of the seminar, 94% consider that the formation has been carried out in Tozeur in the best conditions. However, the opinions vary concerning the personal preparation before reaching Tozeur.

Concerning this issue, it is necessary to draw suggestions in order to improve the service to provide to the participants. It is important to remark that before the seminar the participants did not use absolutely the web site prepared just for them in order to share the information.

5 Recommendations

The seminar "Tozeur 2006" has left a positive impact (Cf: 4.2.1). With the purpose of better understanding the phenomenon of the desertification and to build a better cooperation between the two coasts of the Mediterranean, the participants have expressed their desire to see establishing in the region of Tozeur and under the aegis of IRA/UNICA and NatoSci a continuous seminar that for each proposed formation topic, would analyze correspondent issues (Cf: 4.2.1), with development of courses also in other states. This formation constitutes one manifested urgent necessity from the participants (professionists of the survey of countryside).

5.1 Solidal initiative of search on the development of the barren zones

Based on the outcomes of the formation of Touzer, on the lessons drawn from the tunisian experience and on the necessities expressed by the participants, the organizers and the entertainers have carried out numerous meetings in order to estimate the job and to delineate the lines of cooperation for the future.

As a result of the profitable exchanges of opinion, the participants appreciate the patronage of the NatoSci Program that has allowed to re-unite specialists of numerous countries and various specialties. They also appreciate the good conditions in which the seminar of Tozeur has been carried out and the recommendations that followed.

The organization of moving seminars concerning the different topics of the protection, the conservation and the reorganization of the resources of the barren environment is a priority in the framework of the strengthening of the countries that are touched by desertification.

For that purpose the organizers will stimulate the Program Nato Sci to analyze the possibilities of realization of that formation.

The organizers also consider that the South of Tunisia (and Touzer in particular), because of the scientist available, the logistic and the originality of the efforts undertaken, is an ideal place for a formation that conveying participants from the North and the South of the Mediterranean.

In order to push the Inter-Mediterranean cooperation concerning the fight against the desertification, the idea to realize and to introduce for funding a plan of research-development of regional character, having as a common topic the oasis system has been taken in consideration.

Therefore a group of work has been created, which has the task of drawing the outlines of proposals and to examine the possibilities of funding.

6 The appeal of Tozeur for the conservation of the oasis all over the world

At the end of the seminar of high formation and based on the approach of the management of the resources in the oasis of Tozeur and on the importance of the environment-oasis as world-wide heritage for the fight against the desertification, the preservation of the resources and of the "know-how", the participants launch the appeal of Tozeur for the conservation of the oasis in the world.

The appeal aims to develop the necessary guarantees for the safeguard and the long-lasting reorganization of the natural and human environment of the oasis; formerly to the new dangers that threaten the landscapes, the ecology and the quality of the life of the populations of the oases.

7 Scientific content

The first day (scientific sessions 1 and 2), has been dedicated to conferences about the general aspects of the desertification, to the various aspects of the barren landscapes and to their dynamics. A special session has allowed the participants to introduce their present and future works. The third session has been dedicated to the aspects of remote sensing and to the geographic information systems; to the multiple applications of these surveying systems as tools for the fight against the desertification and for the reasoned management of natural resources. Those three sessions have been followed by a general discussion that allowed to reach the following conclusions:

- Desertification is an universal phenomenon, it exceeds the frontiers of the different countries; hundreds million of persons all over the world are directly touched.
- Scientific research has an important role and must be oriented toward the applicative aspects in a strict relation with the demands of development.
- A reliable information in constant modernization is a necessity and a guarantee for the long-lasting management of the natural resources.
- The techniques of survey, particularly remote sensing, as well as geomatic, constitute privileged ways in order to control the desertification and to support the necessary decisions.

Indeed, these techniques allow measuring the impact of the phenomenon on the people and on the natural environment.

An interdependent and international partnership has primary importance, particularly in the field of the inventory, and in the control and the evaluation of the land deterioration in barren zones.

8 Acknowledgement

At the closure of the seminar of Tozeur, the participants want to express their thanks to:

- The Mayor of Tozeur, who is both an artist and man of letters, for all the support that he has not stopped to supply to the organizers and the participants
- The Tunisian authorities (Gouvernorat di Tozeur)
- The Institute for the barren regions in Medenine
- The University of Cagliari with the TeleGIS Laboratory and the Interdepartmental Center of Engineering and Environmental Science
- The city and the population of Tozeur
- The regional Commissioner for the agricultural development and his employees

- The group for the development of biodynamic agriculture (GDABD)
- The center of research for the Cultures Oasiennes in Dégache
- The staff of the hotel and of the Musée Dar Cherait and Parc ChakWak, for the quality of the acceptance during all the period of the seminar in Tozeur
- El Haj Abdallah Gasmi for giving the best picture of the local cuisine
- Mr. Slim of TOST agency and his nice group of drivers of the cross-country vehicles

The participants also express their admiration and their thankfulness to the efforts of the city administration, that developed the conditions in Touzer, making it a unique Saharian city.

Also thank the Tunisian television, Gafsa radio and the Tunisian newspapers for their interest towards the seminar of Tozeur.

TOZEUR DECLARATION

To preserve the environment and the oasis landscape

Pictures out of the chapters are by Giacomo Marini

Printed in the United States
131824LV00002B/118/P